Designing of a PV/Wind/Diesel Hybrid Energy System
By the aid of the Micro-Grid Modelling Software HOMER Pro® of NREL

Copyright © 2017 by Ali Mubarak

All rights reserved.

Without limiting the rights under copyright reserved above, no part of this publication may be reproduced, stored in or introduced into a retrieval system, or transmitted, in any form, or by any means (electronic, mechanical, photocopying, recording, or otherwise) without the prior written permission of the copyright owner of this book.

ISBN: 978-1-365-94520-5

Preface

PV/Wind/Diesel Hybrid Energy Systems are used for energy supply in remote areas. Depending on the geographical and meteorological situation as well as on the load profile specific hybrid energy system designs are the best and most cost efficient solution. Typically, these systems are used in regions where the costs for a grid connection are too high due to the remote necessaries. The advantage of these systems is that it is possible to design them very flexible and adjust them to the concrete demands. On the other hand, they represent a complex system thus requiring some effort in planning and design.

The proposed system is intended to supply a small living complex in a remote area in Khartoum (Sudan). The aim is to find the best system configuration and to what extent the combination of these two renewable energy sources can reduce the costs compared to a grid-connected system.

The system has been evaluated and optimized with the assistance of the simulation tool HOMER developed by the National Renewable Energy Laboratory (NREL), Colorado, USA.

Nomenclatures and Abbreviations

AC = Alternating current
a-Si = Amorphous Silicon
CdTe = Cadmium Telluride
CIS = Copper Indium Diselenide
DC = Direct Current
DOD = Depth-Of-Discharge
EMI = Electromagnetic Interference
ERI = Energy Research Institute, Sudan
FET = Field Effect Transistor
GEF = Global Environment Facility, USA
HAWT = Horizontal-Axis Wind Turbine
HES = Hybrid Energy Systems
HOMER = The micropower optimization model of NREL, USA
IGBT = Insulating Gate Bipolar Transistor
I_L = Load Current
I_{PV} = PV Current
I_{SC} = Short-Circuit current
kWh/kWp = kilowatt-hours per kilowatt peak
MOSFET = Metal Oxide Semiconductor Field Effect Transistor
MPPT = Maximum Power Point Tracking
NEC = National Electric Codes, USA
NREL = National Renewable Energy Laboratory, USA
PSH = Peak Solar Hours
PV = Photovoltaic
PWM = Pulse Width Modulation
RAPS = Remote Area Power Supplies
RE = Renewable Energy
RES = Renewable Energy System
SOC = State-Of-Charge
STC = standard test conditions
TOE = Tons of Oil Equivalent
UL = Underwriter Laboratories, USA
UNDP = United Nations Development Programme
VAWT = Vertical-Axis Wind Turbine
WG = Wind Generator

CONTENTS

Copyright Page i
Preface ii
Nomenclatures iii
Contents iv

1. INTRODUCTION
  1.1 Renewable Energy and Climate Change.................................. 1
  1.2 Renewable Energy sources 8
    1.2.1 Solar Energy... 8
    1.2.2 Wind Energy... 10
    1.2.3 Hydropower Energy... 11
    1.2.4 Biomass Energy.. 14
    1.2.5 Wave Energy.. 15
    1.2.6 Tidal Energy... 17
    1.2.7 Hydrogen Energy... 18
    1.2.8 Geothermal Energy... 22
  1.3 The status and applications of solar and wind energies in Sudan...... 23
    1.3.1 Solar energy potentials and development 24
    1.3.2 Wind energy potentials and development 27

2. Renewable Systems: Components and Types
  2.1 Components of renewable energy system............................. 30
    2.1.1 Photovoltaic Cells.. 30
      2.1.1.1 What is a Photovoltaic cell? 30
      2.1.1.2 Photovoltaics technology.. 34
    2.1.2 Wind Turbines 39
      2.1.2.1 Components of a wind turbine................................. 39
      2.1.2.2 Turbine design... 41
    2.1.3 Storage batteries 43
    2.1.4 Power Conditioning and Control 46
      2.1.4.1 PV charge controllers.. 46
      2.1.4.2 DC-DC converter.. 49
      2.1.4.3 Inverters... 50
      2.1.4.4 Wind turbine charge controllers.............................. 53
    2.1.5 Back-up generator... 53
  2.2 Types of renewable energy systems....................................... 55

2.2.1 Stand-alone systems	55
2.2.2 Grid-connected systems	55
2.2.3 Hybrid systems	56
2.2.3.1 Series hybrid energy system	58
2.2.3.2 Switched hybrid energy system	61
2.2.3.3 Parallel hybrid energy system	63
3. Sizing the hybrid system	
3.1 Introduction	68
3.2 Sun and Wind data and load demand profile	69
3.2.1 Sun and Wind data	69
3.2.2 Load demand profile	72
3.3 Inverter sizing	74
3.4 Battery sizing	74
3.5 Determining the number of parallel and serial PV modules	76
3.6 Sizing the charge regulator	79
3.7 Choosing the wind turbine and the generator	79
3.8 HOMER simulation and optimization tool	81
3.8.1 What's HOMER?	81
3.8.2 Program Inputs	84
3.8.2.1 Solar and PV inputs	84
3.8.2.2 Wind and turbine inputs	87
3.8.2.3 Diesel generator inputs	87
3.8.2.4 Load profile	88
3.8.3 Program outputs and results	89
4. Conclusion	102
5. References	

Chapter 1: Introduction

1.1 Renewable Energy and Climate Change

Today, the world is facing an impending energy crisis and meanwhile we are also facing the environmental crisis caused by climate change and greenhouse/polluting gas emissions. Development of renewable (or alternative) energy technologies will not only make energy independence feasible, but it will protect our Earth home and provide healthier environments for human beings. Nowadays, people from relevant fields are bringing a broad range of expertise to radically increase the utilization of renewable energy and alternative fuels. Most renewable energy comes directly or indirectly from the sun, thus, the energy resource will not be depleted in the foreseeable future. Furthermore, the energy security of a country can be significantly enhanced by fully utilizing renewable energy due to its decreased reliance on imported fossil fuels [01].

Renewable energy sources are expected to become economically competitive as their costs already have fallen significantly compared with conventional energy sources in the medium term, especially if the massive subsidies to nuclear and fossil forms of energy are phased out. Finally, new renewable energy sources offer huge benefits to developing countries, especially in the provision of energy services to the people who currently lack them. Up to now, the renewable sources have been completely discriminated against for economic reasons. However, the trend in recent years favors the renewable sources in many cases over conventional sources.

The advantages of renewable energy are that they are sustainable (non-depletable), ubiquitous (found everywhere across the world in contrast to fossil fuels and minerals), and essentially clean and environmentally friendly. The disadvantages of renewable energy are its variability, low density, and generally higher initial cost. For different forms of renewable energy, other disadvantages or perceived problems are pollution, odor from biomass, avian with wind plants, and brine from geothermal.

In contrast, fossil fuels are stored solar energy from past geological ages. Even though the quantities of oil, natural gas, and coal are large, they are finite and for the long term of hundreds of years they are not sustainable. The world energy demand depends, mainly, on fossil fuels with respective shares of petroleum, coal, and natural gas at 38%, 30%, and 20%, respectively. The remaining 12% is filled by the non-conventional energy alternatives of hydropower (7%) and nuclear energy (5%).

It is expected that the world oil and natural gas reserves will last for several decades, but the coal reserves will sustain the energy requirements for a few centuries.

This means that the fossil fuel amount is currently limited and even though new reserves might be found in the future, they will still remain limited and the rate of energy demand increase in the world will require exploitation of other renewable alternatives at ever increasing rates. The desire to use renewable energy sources is not only due to their availability in many parts of the world, but also, more empathetically, as a result of the fossil fuel damage to environmental and atmospheric cleanness issues. The search for new alternative energy systems has increased greatly in the last few decades for the following reasons:

1. The extra demand on energy within the next five decades will continue to increase in such a manner that the use of fossil fuels will not be sufficient, and therefore, the deficit in the energy supply will be covered by additional energy production and discoveries.

2. Fossil fuels are not available in every country because they are unevenly distributed over the world, but renewable energies, and especially solar radiation, are more evenly distributed and, consequently, each country will do its best to research and develop their own national energy harvest.

3. Fossil fuel combustion leads to some undesirable effects such as atmospheric pollution because of the CO_2

emissions and environmental problems including air pollution, acid rain, greenhouse effect, climate changes, oil spills, etc. It is understood by now that even with refined precautions and technology, these undesirable effects can never be avoided completely but can be minimized. One way of such minimization is to substitute at least a significant part of the fossil fuel usage by solar energy.

In fact, the worldwide environmental problems resulting from the use of fossil fuels are the most compelling reasons for the present vigorous search for future alternative energy options that are renewable and environmentally friendly. The renewable sources have also some disadvantages, such as being available intermittently as in the case of solar and wind sources or fixed to certain locations including hydropower, geothermal, and biomass alternatives. Another shortcoming, for the time being, is their transportation directly as a fuel. These shortcomings point to the need for intermediary energy systems to form the link between their production site and the consumer location, as already mentioned above. If, for example, heat and electricity from solar power plants are to be made available at all times to meet the demand profile for useful energy, then an energy carrier is necessary with storage capabilities over long periods of time for use when solar radiation is not available.

The use of conventional energy resources will not be able to offset the energy demand in the next decades but steady increase will continue with undesirable environmental consequences. However, newly emerging renewable alternative energy resources are expected to take an increasing role in the energy scenarios of the future energy consumptions [02].

According to the data in (Table 1.1) 13.3% of the world's total primary energy supply came from RE in 2003. However, almost 80% of the RE supply was from biomass (Figure 1.1) and in developing countries it is mostly converted by traditional open combustion, which is very inefficient. Because of its inefficient use, biomass resources presently supply only about 20% of what they could if converted by more efficient, already available technologies.

TABLE 1.3   2003 Fuel Shares in World Total Primary Energy Supply

Source	Share (%)
Oil	34.4
Natural Gas	21.2
Coal	24.4
Nuclear	6.5
Renewables	13.3

Source: Data from IEA, *World Energy Outlook*, IEA, Paris, 2004. With permission.

Source: Reference [03]

(Table 1.1) 2003 Fuel Shares in World Total Primary Energy Supply

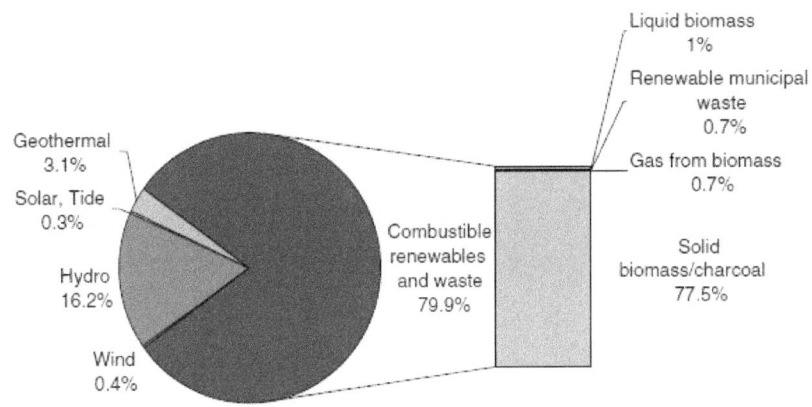

Source: Reference [03]

(Figure 1.1) 2003 resource shares in world renewable energy supply

As it stands, biomass provides only 11% of the world total primary energy, which is much less than its real potential. The total technologically sustainable biomass energy potential for the world is 3–4 TWe, which is more than the entire present global electrical generating capacity of about 3 TWe. In 2003, shares of biomass and hydropower in the total primary energy mix of the world were about 11% and 2%, respectively. All of the other renewables, including solar thermal, solar PV, wind, geothermal, and ocean combined, provided only about 0.5% of the total primary energy. During the same year, biomass combined with hydroelectric resources provided more than 50% of all the primary energy in Africa, 29.2% in Latin America, and 32.7% in Asia (Table 1.2). However, biomass is used very inefficiently for cooking in these countries. Such use has also resulted in significant health problems, especially for women. The total share of all renewables for electricity production in 2002 was about 17%, a vast majority (89%) of it being from hydroelectric power (Table 1.3). [3]

Region	TPES	Renewables	Share of Renewables in TPES (%)
	MTOE	MTOE	
Africa	558.9	279.9	50.1
Latin America	463.9	135.5	29.2
Asia	1224.4	400	32.7
India	553.4	218	39.4
China	1425.9	243.4	17.1
Non-OECD[a] Europe	103.5	9.7	9.4
Former U.S.S.R.	961.7	27.5	2.9
Middle East	445.7	3.2	0.7
OECD	5394.7	304.7	5.6
U.S.A.	2280.8	95.3	4.2
World	10,578.7	1403.7	13.3

[a] OECD, Organization for Economic Cooperation and Development.
Source: From IEA, *Renewables Information*, IEA, Paris, 2005. With permission.

Source: Reference [03]

(Table 1.2) Share of Renewable Energy (RE) in 2003 Total Primary Energy Supply (TPES) on a Regional Basis

Energy Source	2002	
	TWh	(%)
Hydropower	2610	89
Biomass	207	7
Wind	52	2
Geothermal	57	2
Solar	1	0
Tide/Wave	1	0
Total	2927	100

Source: Data from IEA, *World Energy Outlook*, IEA, Paris, 2004. With permission.

Source: Reference [03]

(Table 1.3) Electricity from RE in 2002

1.2 Renewable Energy Resources

In this section, the characteristics of several renewable sources of energy currently utilized in the world are discussed and I will then focus in this study on the solar (Photovoltaic) and wind energies; their applications and their energy systems' components.

1.2.1 Solar Energy

Almost all the renewable energy sources originate entirely from the sun. The sun's rays that reach the outer atmosphere are subjected to absorption, reflection, and transmission processes through the atmosphere before reaching the earth's surface. On the other hand, depending on the earth's surface topography, as explained by Neuwirth (1980), the solar radiation shows different appearances.

The emergence of interest in solar energy utilization has taken place since 1970, principally due to the then rising cost of energy from conventional sources. Solar radiation is the world's most abundant and permanent energy source. The amount of solar energy received by the surface of the earth per minute is greater than the energy utilization by the entire population in one year. For the time being, solar energy, being available everywhere, is attractive for stand-alone systems particularly in the rural parts of developing nations. Occurrences of solar energy dynamically all over the world in the forms of wind, wave, and hydropower through the hydrological cycle provide abilities to ponder about their utilization, if possible instantly or in the form of reserves by various conversion facilities and technologies. It is also possible that in the very long term, human beings might search for the conversion of ocean currents and temperature differences into appreciable quantities of energy so that the very end product of solar radiation on the earth will be useful for sustainable development.

The design of many technical apparatuses such as coolers, heaters, and solar energy electricity generators in the form of photovoltaic cells, requires terrestrial irradiation data at the study area. Scientific and technological studies in the last three decades tried to convert the continuity of solar energy into sustainability for the human comfort. Accurate estimations of global solar radiation need meteorological, geographic, and astronomical data and especially, many estimation models are based on the easily measurable sunshine duration at a set of meteorology stations.

Solar energy is referred to as renewable and/or sustainable energy because it will be available as long as the sun continues to shine. Estimates for the life of the main stage of the sun are another 4 – 5 billion years. The energy from the sunshine, electromagnetic radiation, is referred to as insolation.

Wind energy is derived from the uneven heating of the earth's surface due to more heat input at the equator with the accompanying transfer of water by evaporation and rain. In this sense, rivers and dams for hydro-electric energy are stored solar energy. The third major aspect of solar energy is its conversion into biomass by photosynthesis. Animal products such as whale oil and biogas from manure are derived from solar energy.

1.2.2 Wind Energy

It is one of the most significant and rapidly developing renewable energy sources all over the world. Recent technological developments, fossil fuel usage, environmental effects, and the continuous increase in the conventional energy resources have reduced wind energy costs to economically attractive levels, and consequently, wind energy farms are being considered as an alternative energy source in many enterprises. Although the amount of wind energy is economically insignificant for the time being in many parts of the world, mankind has taken advantage of its utilization for many centuries whenever human beings found the chance to provide power for various tasks. Among these early utilizations are the hauling of water from a lower to a higher elevation, grinding grains in mills by water and other mechanical power applications. It is still possible to see in some parts of the world these types of marginal benefits from wind speed. All previous activities have been technological and the scientific investigation of wind power formulations and accordingly development of modern technology appeared after the turn of the twentieth century. In recent decades the significance of wind energy has originated from its friendly behavior to the environment so far as air pollution is concerned, although there is, to some extent, noise and appearance pollution from the modern wind farms. Due to its cleanness, wind power is sought wherever possible for conversion to electricity with the hope that the air pollution as a result of fossil fuel burning will be reduced. In some parts of the USA, up to 20% of electrical power is generated from wind energy. After the economic crisis of 1973 its importance increased due to economic factors and today there are wind farms in many western European countries. Although the technology in converter-turbines for harnessing the wind energy is advancing rapidly, there is a need to assess its accurate behavior with scientific approaches and calculations.

Wind power is now a reliable and established technology which is able to produce electricity at costs competitive with coal and nuclear power. There will be a small increase in the annual wind energy resource over the Atlantic and northern Europe, with more substantial increases during the winters by 2071 to 2100.

1.2.3 Hydropower Energy

Hydropower is an already established technological way of renewable energy generation. In the industrial and surface water rich countries, the full-scale development of hydroelectric energy generation by turbines at large-scale dams is already exploited to the full limit, and consequently, smaller hydro systems are of interest in order to gain access to the marginal resources. The world's total annual rainfall is, on average, 108.4×10^{12} tons/year of which 12×10^{12} tons recharges the groundwater resources in the aquifers, 25.13×10^{12} tons appears as surface runoff, and 71.27×10^{12} tons evaporates into atmosphere. If the above rainfall amount falls from a height of 1000m above the earth surface, then kinetic energy of 1.062×10^{15} kJ is imparted to the earth every year. Some of this huge amount of energy is stored in dams, which confine the potential energy so that it can be utilized to generate hydroelectric power. Wilbanks (2007) stated that hydropower generation is likely to be impacted because it is sensitive to the amount, timing, and geographical pattern of precipitation as well as temperature (rain or snow, timing of melting). Reduced stream flows are expected to jeopardize hydropower production in some areas, whereas greater stream flows, depending on their timing, might be beneficial. According to Breslow and Sailor (2002), climate variability and long-term climate change should be considered in siting wind power facilities.

As a result of climate change by the 2070s, hydropower potential for the whole of Europe is expected to decline by 6%, translated into a 20 – 50% decrease around the Mediterranean, a 15 – 30% increase in northern and Eastern Europe, and a stable hydropower pattern for western and central Europe. Another possible conflict between adaptation and mitigation might arise over

water resources. One obvious mitigation option is to shift to energy sources with low greenhouse gas emissions such as small hydropower. In regions, where hydropower potentials are still available, and also depending on the current and future water balance, this would increase the competition for water, especially if irrigation might be a feasible strategy to cope with climate change impacts on agriculture and the demand for cooling water by the power sector is also significant. This reconfirms the importance of integrated land and water management strategies to ensure the optimal allocation of scarce natural resources (land, water) and economic investments in climate change adaptation and mitigation and in fostering sustainable development. Hydropower leads to the key area of mitigation, energy sources and supply, and energy use in various economic sectors beyond land use, agriculture, and forestry.
The largest amount of construction work to counterbalance climate change impacts will be in water management and in coastal zones. The former involves hard measures in flood protection (dykes, dams, flood control reservoirs) and in coping with seasonal variations (storage reservoirs and inter-basin diversions), while the latter comprises coastal defense systems (embankment, dams, storm surge barriers). Adaptation to changing hydrological regimes and water availability will also require continuous additional energy input. In water-scarce regions, the increasing reuse of waste water and the associated treatment, deep-well pumping, and

especially large-scale desalination, would increase energy use in the water sector.

1.2.4 Biomass Energy

Overall 14% of the world's energy comes from biomass, primarily wood and charcoal, but also crop residue and even animal dung for cooking and some heating. This contributes to deforestation and the loss of topsoil in developing countries. Biofuel production is largely determined by the supply of moisture and the length of the growing season (Olesen and Bindi 2002). By the twenty-second century, land area devoted to biofuels may increase by a factor of two to three in all parts of Europe (Metzger et al., 2004). Especially, in developing countries biomass is the major component of the national energy supply. Although biomass sources are widely available, they have low conversion efficiencies. This energy source is used especially for cooking and comfort and by burning it provides heat. The sun's radiation that conveys energy is exploited by the plants through photosynthesis, and consequently, even the remnants of plants are potential energy sources because they conserve historic solar energy until they perish either naturally after very long time spans or artificially by human beings or occasionally by forest fires. Only 0.1% of the solar incident energy is used by the photosynthesis process, but even this amount is ten times greater than the present day world energy consumption. Currently, living plants or remnants from the past are reservoirs of biomass that are a major source of energy for humanity in the future. However, biomass energy returns its energy to the atmosphere partly by respiration of the living plants and partly by oxidation of the carbon fixed by photosynthesis that is used to form fossil sediments which eventually transform to the fossil fuel types such as coal, oil, and natural gas. This argument shows that the living plants are the recipient media of incident solar radiation and they give rise to various types of fossil fuels. Biofuel crops, increasingly an important source of energy, are being assessed for their critical role in adaptation to climatic change and mitigation of carbon emissions.

1.2.5 Wave Energy

Water covers almost two thirds of the earth, and thus, a large part of the sun's radiant energy that is not reflected back into space is absorbed by the water of the oceans.

This absorbed energy warms the water, which, in turn, warms the air above and forms air currents caused by the differences in air temperature. These currents blow across the water, returning some energy to the water by generating wind waves, which travel across the oceans until they reach land where their remaining energy is expended on the shore.

The possibility of extracting energy from ocean waves has intrigued people for centuries. Although there are a few concepts over 100 years old, it is only in the past two decades that viable schemes have been proposed. Wave power generation is not a widely employed technology, and no commercial wave farm has yet been established.

In the basic studies as well as in the design stages of a wave energy plant, the knowledge of the statistical characteristics of the local wave climate is essential, no matter whether physical or theoretical/numerical modeling methods are to be employed. This information may result from wave measurements, more or less sophisticated forecast models, or a combination of both, and usually takes the form of a set of representative sea states, each characterized by its frequency of occurrence and by a spectral distribution. Assessment of how turbo-generator design and the production of electrical energy are affected by the wave climate is very important. However, this may have a major economic impact, since if the equipment design is very much dependent on the wave climate, a new design has to be developed for each new site. This introduces extra costs and significantly limits the use of serial construction and fabrication methods.

Waves have an important effect in the planning and design of harbors, waterways, shore protection measures, coastal structures, and the other coastal works. Surface waves generally derive their energy from the wind. Waves in the ocean often have irregular shapes and variable propagation directions because they are under the influence of the wind. For operational studies, it is desired to forecast wave parameters in advance. Özger and ‚Sen (2005) derived a modified average wave power formula by using perturbation methodology and a stochastic approach.

1.2.6 Tidal Energy

Another possibility of obtaining energy from the ocean is tidal energy, which exploits the different water level between high tide and low tide. The period between two high or low tides is 12.4 h. The tides are caused by gravitational attraction of the sun and moon.

Large tides can be observed in certain coastal formations like narrow inlets or funnel-shaped bays. In such places, the tides can reach 20 m. The tidal power is harnessed by damming the outlets with dams containing reversible turbines. Thus, both directions of the tide can be used. Some tidal power plants exist, for instance, in St. Malo at the mouth of the Rance River in France (240MW, erected between 1961 and 1967) and at the bay of Fundy in Canada (20MW, erected in 1984).

The potential of tidal power is limited, because it is restricted to certain coastal formations. In addition, such plants should not have too much environmental impact. Under consideration of the above mentioned criteria there exist worldwide only forty to fifty potential locations for tidal power plants with a capacity of 200MW each. At some of those locations, a capacity of more than 1,000MW should be possible. The total potential capacity is, therefore, 10GW. The development of tidal energy has been very slow considering the fact that the pilot plants have existed for decades and the technology is basically known.

1.2.7 Hydrogen Energy

Hydrogen is the most abundant element on earth, however, less than 1% is present as molecular hydrogen gas H_2; the overwhelming part is chemically bound as H_2O in water and some is bound to liquid or gaseous hydrocarbons. It is thought that the heavy elements were, and still are, being built from hydrogen and helium. It has been estimated that hydrogen makes up more than 90% of all the atoms or 75% of the mass of the universe. Combined with oxygen it generates water, and with carbon it makes different compounds such as methane, coal, and petroleum.

Hydrogen exhibits the highest heating value of all chemical fuels. Furthermore, it is regenerative and environment friendly.

Solar radiation is abundant and its use is becoming more economic, but it is not harvested on large scale. This is due to the fact that it is difficult to store and move energy from ephemeral and intermittent sources such as the sun. In contrast, fossil fuels can be transported easily from remote areas to the exploitation sites. For the transportation of electric power, it is necessary to invest and currently spend money in large amounts. Under these circumstances of economic limitations, it is more rational to convert solar power to a gaseous form that is far cheaper to transport and easy to store. For this purpose, hydrogen is an almost completely clean-burning gas that can be used in place of petroleum, coal, or natural gas. Hydrogen does not release the carbon compounds that lead to global warming. In order to produce hydrogen, it is possible to run an electric current through water and this conversion process is known as electrolysis. After the production of hydrogen, it can be transported for any distance with virtually no energy loss. Transportation of gases such as hydrogen is less risky than any other form of energy, for instance, oil which is frequently spilled in tanker accidents, or during routine handling.

The ideal intermediary energy carrier should be storable, transportable, pollution-free, independent of primary resources, renewable, and applicable in many ways. These properties may be met by hydrogen when produced electrolytically using solar radiation, and hence, such a combination is referred to as the solar-hydrogen process. This is to say that transformation to hydrogen is one of the most promising methods of storing and transporting solar energy in large quantities and over long distances.

Among the many renewable energy alternatives, solar-hydrogen energy is regarded as the most ideal energy resource that can be exploited in the foreseeable future in large quantities. On the other hand, where conventional fuel sources are not available, especially in rural areas, solar energy can be used directly or indirectly by the transformation into hydrogen gas. The most important property of hydrogen is that it is the cleanest fuel, being non-toxic with virtually no environmental problems during its production, storage, and transportation. Combustion of hydrogen with oxygen produces virtually no pollution, except its combustion in air produces small amounts of nitrogen oxides. Solar-hydrogen energy through the use of hydrogen does not give rise to acid rain, greenhouse effects, ozone layer depletions, leaks, or spillages. It is possible to regard hydrogen after the treatment of water by solar energy as a synthetic fuel. In order to benefit from the unique properties of hydrogen, it must be produced by the use of a renewable source so that there will be no limitation or environmental pollution in the long run. Different methods have been evoked by using direct or indirect forms of solar energy for hydrogen production.

These methods can be viewed under four different processes, namely:
- Direct thermal decomposition or thermolysis
- Thermo-chemical processes
- Electrolysis
- Photolysis

Large-scale hydrogen production has been obtained so far from the water electrolysis method, which can be used effectively in combination with photovoltaic cells. Hydrogen can be extracted directly from water by photolysis using solar radiation.

Photolysis can be accomplished by photobiological systems, photochemical assemblies, or photoelectrochemical cells.

Hydrogen has been considered by many industrial countries as an environmentally clean energy source. In order to make further developments in the environmentally friendly solar-hydrogen energy source enhancement and research, the following main points must be considered:

- It is necessary to invest in the research and development of hydrogen energy technologies
- The technology should be made widely known
- Appropriate industries should be established
- A durable and environmentally compatible energy system based on the solar hydrogen process should be initiated Veziroˇglu (1995) has suggested the following research points need to be addressed in the future to improve the prospects of solar-hydrogen energy:

1. Hydrogen production techniques coupled with solar and wind energy sources
2. Hydrogen transportation facilities through pipelines
3. Establishment and maintenance of hydrogen storage techniques
4. Development of hydrogen-fueled vehicles such as busses, trucks, cars, etc.
5. Fuel cell applications for decentralized power generation and vehicles
6. Research and development on hydrogen hydrides for hydrogen storage and for air conditioning

7. Infrastructure development for solar-hydrogen energy
8. Economic considerations in any mass production
9. Environmental protection studies

On the other hand, possible demonstrations and/or pilot projects include the following alternatives:
1. Photovoltaic hydrogen production facility
2. Hydrogen production plants by wind farms
3. Hydro power plant with hydrogen off-peak generators
4. Hydrogen community
5. Hydrogen house
6. Hydrogen-powered vehicles

In order to achieve these goals, it is a prerequisite to have a data bank on the hydrogen energy industry, its products, specifications, and prices.

1.2.8 Geothermal Energy

All renewable energy sources mentioned so far (with the exception of tidal energy) are derived from solar radiation. A major potential energy resource not related to solar energy is geothermal energy. This energy originates from the natural cooling of the planet and from radioactive decay. The temperature of the crust of the earth increases with increasing depth. It varies depending on geological conditions, but on average it is $1°$ per 33m.

Three types of geothermal resources can be distinguished:

1. High temperature steam resources
Water is heated by geothermal anomalies. Depending on pressure, it can exist either as superheated water or high pressure steam.
2. Warm water resources: These are aquifers with temperatures lower than 100°C. They are not very suitable for electricity generation, but can be very useful for heating.
3. Hot dry rock resources: The hot dry rock technique exploits the heat content of deep rock layers. Whereas the first two resources are relatively close to the surface, hot dry rock heat is tapped several kilometers below the surface. [04]

1.3 The status and applications of Solar and Wind energies in Sudan

Sudan has one of the fastest growing economies in Africa. However, its remotely isolated rural areas pose problems to rural energy management and development because of poor road links with the urban centers and remoteness from the national electrical transmission grid. Development of renewable energy sources, therefore, has a vast potential in Sudan.

Sudan meets approximately 87 % of its energy needs with biomass, while oil supplies 12 %, and the remaining 1 % is produced from hydro and thermal power. The total energy consumed is approximately 11.7 million tons of oil equivalent (TOE), with an estimated 43 % lost in the conversion process. Electricity reaches only about 30% of Sudan's more than 40 M population; this mainly in urban areas. Hence, a major problem for rural people is the inadequate supply of power for lighting, heating, cooking, cooling, water pumping, radio or TV communications and security services. The heavy dependence on biomass threatens the health and future of domestic forests, and the large quantities of oil purchased abroad causes Sudan to suffer from serious trade imbalances. A shift to renewables would therefore help to solve some of these problems while also providing the population with higher quality energy, which will in turn, improve living standards and help reduce poverty. [05]

1.3.1 Solar energy potentials and development

The climate diversity in Sudan (arid in the north and equatorial in the far south), enables Sudan to enjoy vast quantity of solar energy.

The average annual daily solar radiation is 22.84 $MJ/m^2/day$ which is considered one of the highest in the world.

The annual global solar radiation is between 6.87 to 9.75 $MJ/m^2/years$ on horizontal surface. However, the dispersed solar radiation is relatively high in Sudan and reached 20-40% of the global solar radiation, as result of dust storms in summer, this may affect the performance of the solar collectors negatively. Although, sun shine duration reaches 12 hours/day i.e. 4380 hours in the year, its least average solar radiation of 6 KW/m^2/day can secure electricity to more than 5 million households. (Reference-06)

Since the beginning of year 2000, the government of Sudan put one million dollars on solar energy for solar electrification of rural areas. Two projects on solar energy are operating:

1) <u>Solar Energy Project to Remove Barriers:</u>
The main objective of this project is to remove barriers that hinder dissemination of renewable energy technologies in Sudan this includes:

- Commercial dissemination of solar photovoltaics.
- Development of financial, technological structural and public awareness for commercialization of PV technologies.
- Replace the diesel generators with this one for less environmental impact hazards.
- Develop market strategies to meet demand in commercial basis.
- Transfer policies, manufacturing and construction.
  This project was implemented with share co-operation between the Government of Sudan, UNDP and GEF.

2) The 1000 Villages Project:

The target of this national project is to light about 1000 villages, in the 26 states with total investment of US$ 3.3 million. This project was implemented with a joint effort between the Ministry of Energy and Mining and the States during 2000-2003. This project provides services in schools, mosques, health centers, hospitals, clubs, pumping water for drinking and irrigation purpose, and for vaccines refrigerators. [06].

Site	Numbers of PV	Power Systems kW	Cost Million SD
Port Sudan	90	5	19.5
Northern Darfur	120	6	20
Southern Darfur	102	5.1	10
Northern State	140	7	20
Western Kordofan State	188	7.4	24.2
ElGadarif	148	7.4	35
Blue Nile Bridge – Khartoum	32	2.4	10
PV Project	370	20	44.8
Total	1150	59	183.5

Source: Reference-06

(Table 1.4) Government and State Joint Solar PV Projects (2001-2003)

Today, the Sudanese government is actively supporting PV policies. The solar PV project has contributed to enhanced awareness of the social and economic potential of PV power and has boosted activities by the National Energy Committee of the National Assembly to enact a Solar Energy Act. In the annual 2004 national development budget, the parliament passed a resolution exempting PV system components from import duties and the value added tax. The government has further decided to invest in a joint venture with China for a module assembly line. It is expected that the combined effects of tax reduction and local assembly will reduce PV costs by 30–40 percent.

The PV market players in Sudan are optimistic and expect increasing sales in coming years. The government and private businesses are hoping for falling PV costs resulting from proposed PV policies and from manufacturing by local firms. They anticipate increased demand from social institutions and private households as they become fully aware of PV's benefits. [07]

1.3.2 Wind energy potentials and development

The metrological stations in Sudan record the readings of wind energy at a height of 15.2 meter. The average wind speed in the Northern, Eastern, Central and Western part of Sudan is approximately 3 m/sec beside that this data includes the direction of the wind at 15.2 meter above ground surface. The wind density in the north part of Sudan, around Dongola exceeds 400 W/m^2 while in Khartoum area it ranges from 285 to 380 W/m^2. [06]

Forty years ago, wind pumps were very common in central Sudan, but gradually disappeared due to a lack of spare parts and maintenance skills combined with stiff competition from relatively cheap diesel pumps. However, the government has recently begun to recognize the need to reintroduce wind pump technology to reduce the country's dependence on foreign oil. This increases economic security, given high and/or fluctuating oil prices, and it helps to reduce the trade deficit. Using wind power also allows for pumping in rural areas where transportation of oil might be difficult.

In 1985, the Energy Research Institute (ERI), in cooperation with a consulting services firm, started a wind pumps project financed by the Netherlands Ministry of Foreign Affairs. During the 14 months of the project, ten imported wind pumps were installed in the Khartoum area, and one was locally manufactured for demonstration. Results suggested that for some local applications, wind energy is an economically and technically viable option. After the termination of the project, the ERI continued monitoring and testing the performance of the installed pumps, and the consulting firm set out to produce wind pumps for low head pumping applications which could be built in developing countries. So far, two wind pumps have been manufactured locally at a cost of US $2,500 each, and initial test results indicate that the design has room for improvement. The performance of the wind pumps was below expectations, possibly due to a low pump efficiency and high start-up wind speed (3 m/s). The amount of maintenance required was also higher than anticipated (at least once every two months).

For wind pumps to be effective and competitive, it is recommended that further research be carried out to improve overall efficiency and simplicity of the wind systems, and to incorporate locally available materials into the design. Also, quality control guidelines should be established and users trained on how to utilize the pumps more efficiently. [05]

Chapter 2: Renewable Systems: Components and Types

2.1 Components of renewable energy system
2.1.1 Photovoltaic Cells
2.1.1.1 What is a Photovoltaic cell?

PV cells convert sunlight directly into electricity without creating any air or water pollution. PV cells are made of at least two layers of semiconductor material. One layer has a positive charge, the other negative. When light enters the cell, some of the photons from the light are absorbed by the semiconductor atoms, freeing electrons from the cell's negative layer to flow through an external circuit and back into the positive layer. This flow of electrons produces electric current.

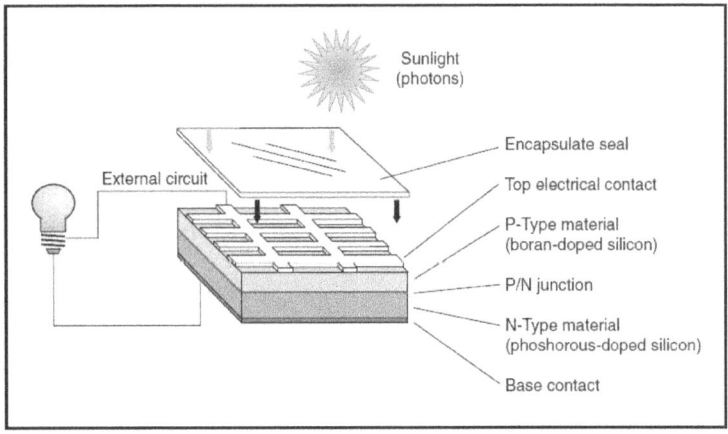

Source: Reference [08]

(Figure 2.1) Basic PV cell construction

To increase their utility, dozens of individual PV cells are interconnected together in a sealed, weather proof package called a module. When two modules are wired together in series, their voltage is doubled while the current stays constant. When two modules are wired in parallel, their current is doubled while the voltage stays constant. To achieve the desired voltage and current, modules are wired in series and parallel into what is called a PV array. The flexibility of the modular PV system allows designers to create solar power systems that can meet a wide variety of electrical needs, no matter how large or small. [08]

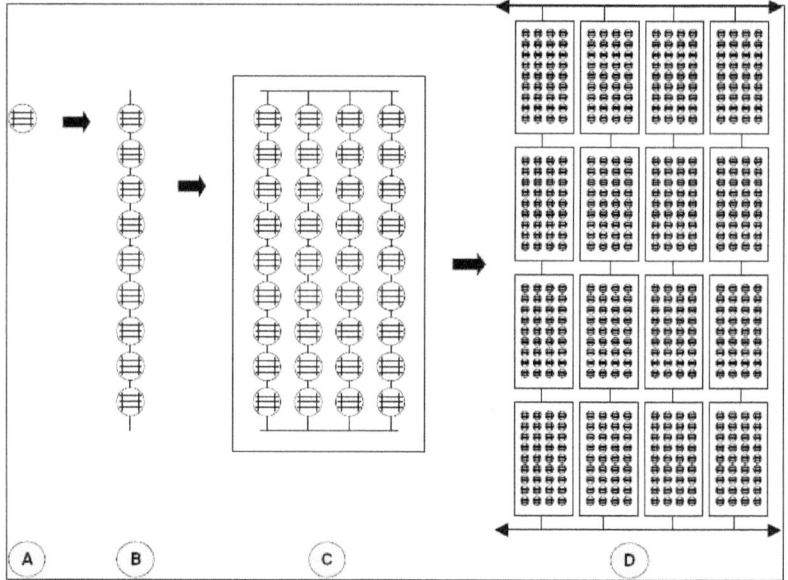

Individual photovoltaic (PV) cells (A) are combined in series strings (B) to obtain higher voltage. Series strings are encapsulated together to form a module (C). Modules are connected in series strings, and module series strings are connected in parallel to form a PV array (D) that produces the desired voltage and power.

Source: Reference [09]

(Figure 2.2) Photovoltaic cells, modules and arrays

Crystalline silicon solar cells comprise approximately 90% of the market share of PV modules. These PV modules typically consist of 36 cells, but other configurations can be bought off the shelf. The current-voltage curve of a 36-cell module fits very well to 12-V lead acid batteries, so simple autonomous power supplies can be built. However, the exact number of 36 cells is more historically motivated. With improved cell technology, even at higher ambient temperatures the module voltage would still be high enough to fully charge a lead acid battery if 32 cells were connected in series. Depending on climatic conditions, even less cells could be used.

The yield of a PV system is usually measured in kilowatt-hours per kilowatt peak (kWh/kWp) and depends on several factors:
- Cell and module technology used
- Location/climatic conditions
- Tilt angle and orientation
- Mechanical installation (facade or roof integration, freely ventilated modules etc., influencing module temperature)
- Efficiency of system components (MPP tracking and conversion efficiency)
- Type of system

The area needed for 1 kW of PV power mainly depends on the cell and module technology used. Module efficiencies vary from several percent (for some thin-film modules) to approximately 16 or 17% for monocrystalline silicon solar cells. Typical module sizes for grid connected systems range from 50 to more than 300W. Especially for solar lanterns and lighting, modules in the range of 10–20 Wp are available. Because 36 cells per module are standard, smaller modules are made from cells that are cut into half or quarter cells. Thus, small PV modules with crystalline silicon solar cells are more expensive. This does not apply to thin-film modules. The performance of PV modules depends on their operation temperature. Depending on the technology used, the temperature coefficient is between -0.2 and -0.5%/K. Thus, the type of installation and degree of rear ventilation are important. PV modules are rated in watt peak under standard test conditions (STCs). The STCs are defined as $1000W/m^2$ insolation with a AM solar spectrum of 1.5 and 251C module temperature. However, these conditions rarely occur in real systems. Thus, PV generators can actually yield significantly less or more energy than rated, depending on the location, climatic conditions, etc. However, there is no widely accepted alternative rating model that takes into account more realistic reporting conditions. Furthermore, there is relatively large uncertainty of the measurements at the manufacturing plant and on site. Spectral mismatch, temperature distribution, etc. comprise approximately 3–5% of uncertainty. Overall deviations in systems measurement can be 5–10%.

Whereas in grid-connected systems usually all energy can be fed into the grid, in autonomous systems part of the energy yield in summer cannot be used due to limited storage capacity. Thus, the tilt angle of PV modules in autonomous systems is usually optimized for large energy yield during winter. The overall yearly energy yield is reduced. [09].

2.1.1.2 Photovoltaics Technology

The photovoltaic effect remained a laboratory curiosity from 1839 until 1959, when the first silicon solar cell was developed at Bell Laboratories in 1954 by Chapin et al. It already had an efficiency of 6%, which was rapidly increased to 10%. The main application for many years was in space vehicle power supplies.
Terrestrial application of photovoltaics (PV) developed very slowly. Nevertheless,
PV fascinated not only the researchers, but also the general public.
Its strong points are:
- Direct conversion of solar radiation into electricity,
- No mechanical moving parts, no noise,
- No high temperatures,
- No pollution,
- PV modules have a very long lifetime,
- The energy source, the sun, is free, ubiquitous, and inexhaustible,

- PV is a very flexible energy source, its power ranging from microwatts to megawatts.

Solar cell technology benefited greatly from the high standard of silicon technology developed originally for transistors and later for integrated circuits. This applied also to the quality and availability of single crystal silicon of high perfection. In the first years, only Czochralski (Cz) grown single crystals were used for solar cells. This material still plays an important role. As the cost of silicon is a significant proportion of the cost of a solar cell, great efforts have been made to reduce these costs. One technology, which dates back to the 1970s, is block casting which avoids the costly pulling process. Silicon is melted and poured into a square SiO/SiN-coated graphite crucible. Controlled cooling produces a polycrystalline silicon block with a large crystal grain structure.

From solid state physics we know that silicon is not the ideal material for photovoltaic conversion. It is a material with relatively low absorption of solar radiation, and, therefore, a thick layer of silicon is required for efficient absorption. Theoretically, this can be explained by the semiconductor band structure of silicon in which the valence band maximum is offset from the conduction band minimum. Since the basic process of light absorption is excitation of an electron from the valence to the conduction band, light absorption is impeded because it requires a change of momentum.

The search for a more suitable material started almost with the beginning of solar cell technology. This search concentrated on the thin-film materials.

They are characterized by a direct band structure, which gives them very strong light absorption. Today, the goal is still elusive, although promising materials and technologies are beginning to emerge. The first material to appear was amorphous Silicon (a-Si). It is remarkable that even the second contender in this field is based on the element silicon, this time in its amorphous form. Amorphous silicon has properties fundamentally different from crystalline silicon. However, it took quite some time before the basic properties of the material were understood. The high expectancy in this material was curbed by the relatively low efficiency obtained so far and by the initial light-induced degradation for this kind of solar cell (so-called Staebler–Wronski effect). Today, a-Si has its fixed place in consumer applications, mainly for indoor use. After understanding and partly solving the problems of light-induced degradation, amorphous silicon begins to enter the power market. Stabilized cell efficiencies reach 13%. Module efficiencies are in the 6–8% range. The visual appearances of thin-film modules make them attractive for façade applications. Beyond amorphous silicon there are many other potential solar cell materials fulfilling the requirement of high light absorption and are therefore suitable for thin-film solar cells. They belong to the class of compound semiconductors like GaAs or InP, which are III–V compounds according to their position in the periodic table. Other important groups are II–VI and I–III–VI2 compounds, which, just like the elemental semiconductors, have four bonds per atom. It is clear that an almost infinite number of compounds could be considered. From the mostly empirical search only very few promising materials have resulted. Foremost are Copper Indium Diselenide (CIS) and Cadmium Telluride (CdTe). Already by the early 1960s cadmium sulfide/copper sulfide solar cells were under development. Problems with low efficiency and insufficient stability prevented further penetration of this material The new technology is based on the ternary compound semiconductors $CuInSe_2$, $CuGaSe_2$, $CuInS_2$ and their multinary alloy $Cu(In,Ga)(S,Se)_2$ (in the following text: CIGS). The first results of single crystal work on $CuInSe_2$ (CIS) were extremely promising, but the complexity of the material looked complicated as a thin-film technology. Pioneering work, however, showed immediate success. It became evident that CIS process technology is very flexible with respect to process conditions. In later developments, the addition of Ga and S helped to increase the efficiency. The best laboratory efficiency has recently reached a remarkable 18.9%. CIS/CIGS modules are now available on the market in small quantities. Thin-film solar cells based on CdTe have a very long tradition and are also just at the onset of commercial production. After a long and varied development phase, they arrived at cell efficiencies of 16% and large-area module efficiencies of over 10%. In spite of the complicated manufacture and the high cost, crystalline silicon still dominates the market today and probably will continue to do so in the immediate future. This is mostly due to the fact that there is an abundant supply of silicon as raw material, high efficiencies are feasible, the ecological impact is low, and silicon in its crystalline form has practically no degradation. The market shares of different technologies in 2002 are shown in (Figure 2.3).

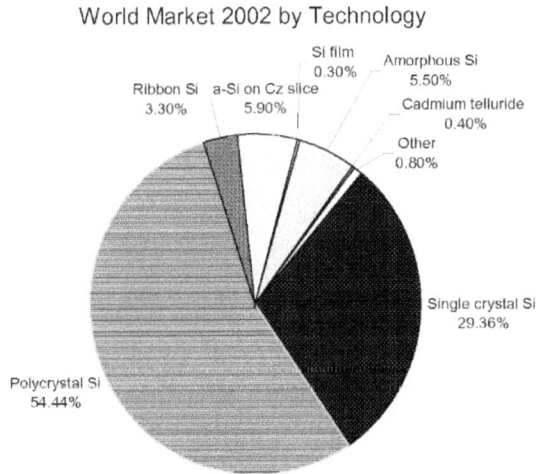

Source: Reference [04]
(Figure 2.3) Market shares of different PV technologies for 2002

The various forms of crystalline silicon have together a share of 93%. Single crystal and cast poly material had about equal share for a long time. Recently, cast

material has surpassed single crystals. Newer types of crystalline silicon like Ribbon and Si film are not yet very important. A newcomer is a-Si on crystalline silicon.
Of the true thin-film materials that are summarized as "others" amorphous silicon is dominant. As mentioned before, its market is mainly in consumer products. These market shares are rather stable and change only in an evolutionary manner. The dominance of the element silicon in its crystalline and amorphous forms is an overwhelming 99%. Of all the other materials only CdTe has a market share of only 0.4%. [04]

2.1.2 Wind Turbines
2.1.2.1 Components of a wind turbine

The wind turbine system consists of blades attached to a central hub that rotate when force is exerted upon them by the wind. The hub is in turn attached to a driveshaft that transmits rotational energy to a generator, housed inside a central enclosure called a nacelle. In most turbine designs, the nacelle must be able to rotate to catch the wind from whatever direction it is coming. The mechanical and electrical controls inside the nacelle include a rotor brake, a mechanical gearbox, a generator, and electrical controls. The yaw mechanism rotates the nacelle relative to the vertical axis of the tower, so that the turbine can face into the wind [10]. Major components of the turbine system are shown in (Figure 2.4)

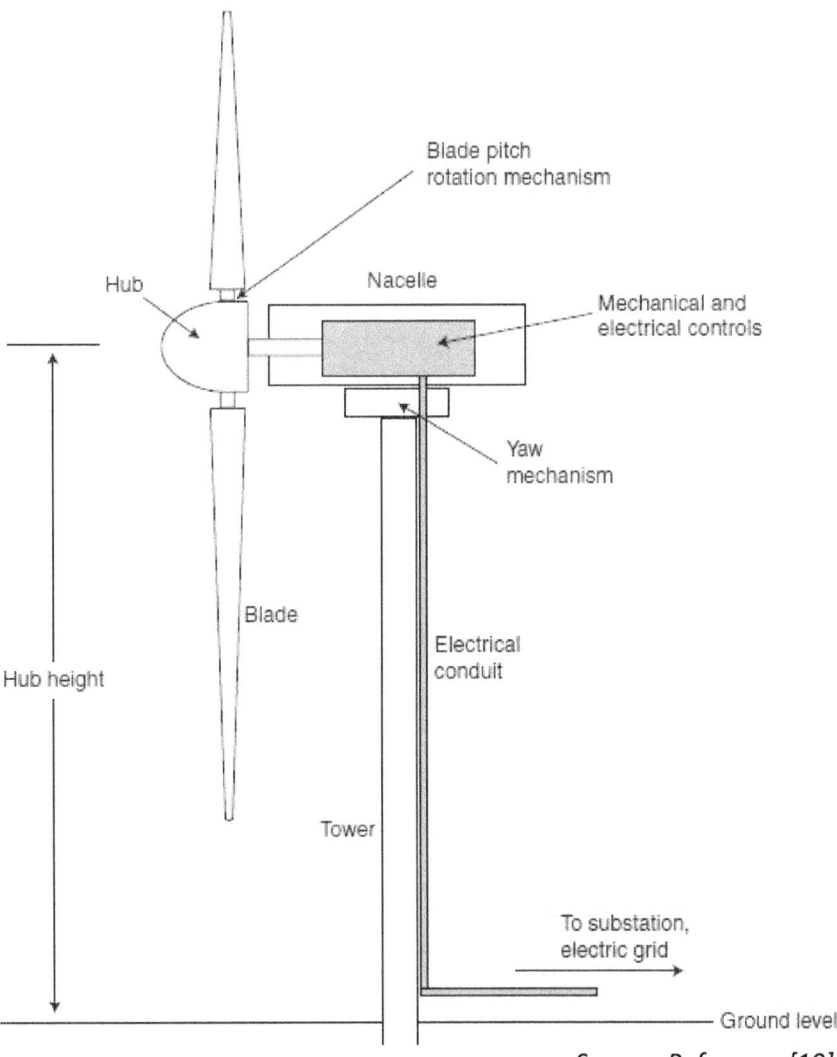

Source: Reference [10]
(Figure 2.4) Main parts of a wind turbine

2.1.2.2 Turbine design

Although there are many different configurations of wind turbines, most of them can be classified as either horizontal-axis wind turbines (HAWTs), which have blades that rotate about a horizontal axis parallel to the wind, or vertical-axis wind turbines (VAWTs), which have blades that rotate about a vertical axis. (Figure 2.5) illustrates the main features of these configurations. They both contain the same major components, but the details of those components differ significantly.

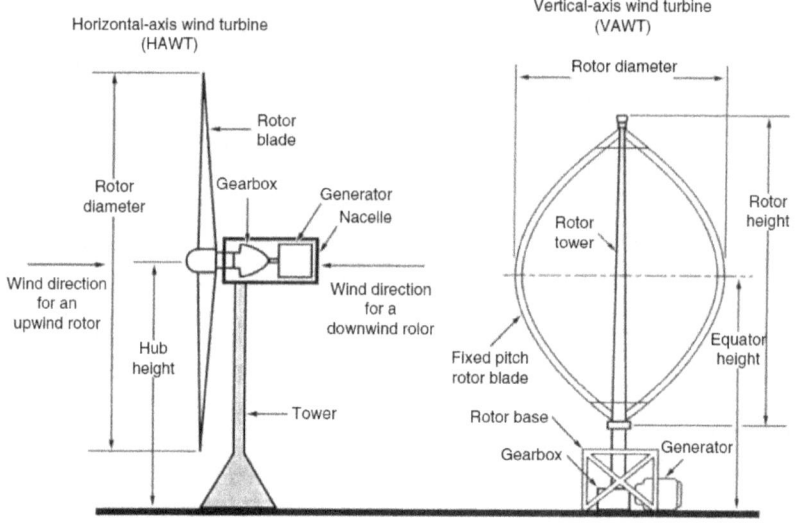

Source: Reference [03]

(Figure 2.5) Schematic of basic wind turbine configurations

As shown in (Figure 2.5), HAWTs and VAWTs have very different configurations. Each configuration has its own set of strengths and weaknesses. HAWTs usually have all of their drive train (the transmission, generator, and any shaft brake) equipment located in a nacelle or enclosure mounted on a tower, as shown. Their blades are subjected to cyclic stresses due to gravity as they rotate, and their rotors must be oriented (yawed) so the blades are properly aligned with respect to the wind. HAWTs may be readily placed on tall towers to access the stronger winds typically found at greater heights. The most common type of modern HAWT is the propeller-type machine, and these machines are generally classified according to the rotor orientation (upwind or downwind of the tower), blade attachment to the main shaft (rigid or hinged), maximum power control method (full or partial-span blade pitch or blade stall), and number of blades (generally two or three blades). VAWTs, on the other hand, usually have most of their drive train on the ground; their blades do not experience cyclic gravitational stresses and do not require orientation with respect to the wind. However, VAWT blades are subject to severe alternating aerodynamic loading due to rotation, and VAWTs cannot readily be placed on tall towers to exploit the stronger winds at greater heights. The most common types of modern VAWTs are the Darrieus turbines, with curved, fixed-pitch blades, and the "H" or "box" turbines with straight fixed-pitch blades. All of these turbines rely on blade stall (loss of lift and increase in drag as the blade angle of attack increases) for maximum power control. Although there are still a few manufacturers of VAWTs, the overwhelming majority of wind turbine manufacturers devote their efforts to developing better (and usually larger) HAWTs [3].

2.1.3 Storage batteries

Systems based on renewable energy resources (PV and Wind) require storage to meet the energy demand during periods of low solar irradiation, nighttime and low wind speeds. Several types of batteries are available such as the lead acid, nickel– cadmium, lithium, zinc bromide, zinc chloride, sodium sulfur, nickel–hydrogen, redox, and vanadium batteries. The provision of cost-effective electrical energy storage remains one of the major challenges for the development of improved Renewable Energy Systems (RES).

Typically, lead-acid batteries are used to guarantee several hours to a few days of energy storage. Their reasonable cost and general availability has resulted in the widespread application of lead-acid batteries for remote area power supplies despite their limited lifetime compared to other system components. Lead-acid batteries can be deep or shallow cycling gelled batteries, batteries with captive or liquid electrolyte, sealed and non-sealed batteries etc. Sealed batteries are valve regulated to permit evolution of excess hydrogen gas (although catalytic converters are used to convert as much evolved hydrogen and oxygen back to water as possible). Sealed batteries need less maintenance. The following factors are considered in the selection of batteries for RES:

- Deep discharge (70–80% depth of discharge).
- Low charging/discharging current.
- Long duration charge (slow) and discharge (long duty cycle).
- Irregular and varying charge/discharge.
- Low self-discharge.
- Long life time.
- Less maintenance requirement.
- High energy storage efficiency.
- Low cost.

Battery manufacturers specify the nominal number of complete charge and discharge cycles as a function of the depth-of-discharge (DOD), as shown in (Figure 2.6). While this information can be used reliably to predict the lifetime of lead-acid batteries in conventional applications, such as uninterruptible power supplies or electric vehicles, it usually results in an overestimation of the useful life of the battery bank in renewable energy systems.

Two of the main factors that have been identified as limiting criteria for the cycle life of batteries in PV power systems are incomplete charging and prolonged operation at a low state-of-charge (SOC). The objective of improved battery control strategies is to extend the lifetime of lead-acid batteries to achieve a typical number of cycles shown in (Figure 2.6). If this is achieved, an optimum solution for the required storage capacity and the maximum DOD of the battery can be found by referring to manufacturer's information. Increasing the capacity will reduce the typical DOD and therefore prolong the battery lifetime. Conversely, it may be more economic to replace a smaller battery bank more frequently [11].

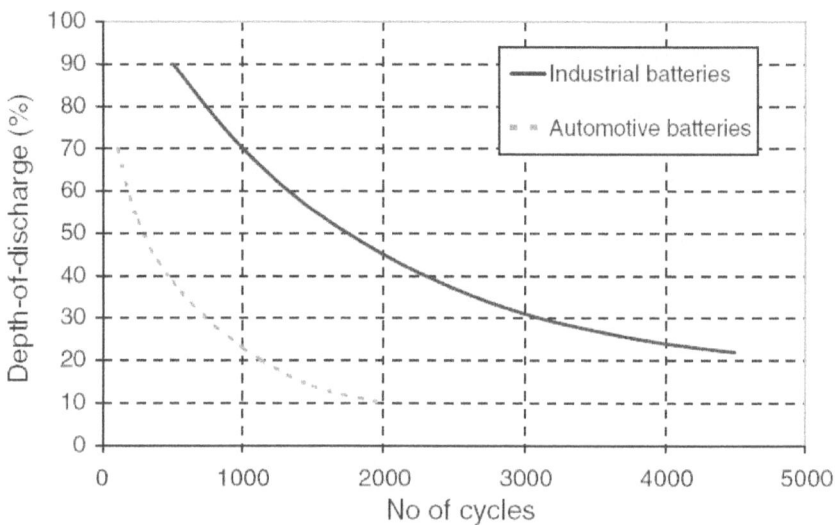

Source: Reference [11]

(Figure 2.6) Nominal number of battery cycles vs DOD

2.1.4 Power conditioning and control
2.1.4.1 PV charge controllers

The role of the charge controller/regulator is to ensure that the battery is not overcharged. There is a bewildering array of commercial devices from a large number of manufacturers, with current ratings from a few amps up to hundreds of amps, and operating voltages generally in multiples of 12 V up to 48 V (nominal battery voltage). Most commercially available charge controllers for PV applications use a switched series regulator to control the charging current (Figure 2.7). The most common control scheme is pulse width modulation (PWM). The power to the battery is switched on and off at a constant frequency, with the duty ratio varied, to control either the mean current to the battery, or the charging voltage of the battery. This scheme is similar to a buck converter, but with a PV power source the current is limited, so a series inductor to limit the peak current is unnecessary, and the load voltage is smoothed by the battery. The control algorithm depends on the type of battery and most charge controllers provide a number of settings to accommodate different voltages and types of lead-acid battery. A typical control scheme would allow continuous current until the battery reaches a predetermined voltage and then the duty ratio of the PWM is reduced to limit the battery voltage. Some controllers use a three stage charging algorithm with a bulk charge phase, where the charging current is the maximum available, a taper phase where the voltage is held constant, and a float phase where the battery voltage is held constant at a reduced value. Many controllers will have control algorithms that will allow different control regimes and/or settings to accommodate different battery types. A few controllers use switched shunt regulation. In this case a transistor switch bypasses the current from the PV array (Figure 2.8). Again PWM may be used to control the mean load current. A variant on the shunt controller uses switches to divert power from charging the battery to a diversion load, again using PWM. The diversion load might be a water heater or similar. This enables the energy not used for charging the battery to be used usefully, rather than dissipated as heat in the PV panel. It is not clear that either series or shunt regulation is to be preferred. Both can achieve high overall efficiency. In both series and shunt controllers the switches used are Field Effect Transistors (FETs) with an ON state resistance of only a few mΩ. Thus the voltage drop across the switch when it is on is usually small, avoiding unnecessary power dissipation, and in the case of a series regulator ensuring that overall efficiency is high. Control schemes other than PWM are possible and are used by some suppliers. One proprietary scheme, Flexcharge™, uses the way in which the battery voltage rises and falls as the current is switched ON and OFF to control the switching. The switch supplies current to charge the battery until the battery voltage reaches a desired upper limit. The current is then switched off until the voltage has fallen to a lower limit. This procedure charges the battery with pulses of current which become shorter and less frequent as the battery approaches full charge. Charge controllers incorporating maximum power point tracking are available, but much less common than simple charge controllers. To be useful a maximum power point tracker must have an efficiency of greater than 90%, this calls for careful design and the cost of such systems is significantly greater than for a basic charge controller. The gain in system efficiency however may be significant; gains of up to 30% are claimed. Such large gains will only be achieved when the battery is in a low state of charge (low voltage) and the PV array is cold (high voltage). Average efficiency gains are however likely to be significantly less (of the order of 10%). The cost of the maximum power point tracker must be weighed against the cost of extra array area [12].

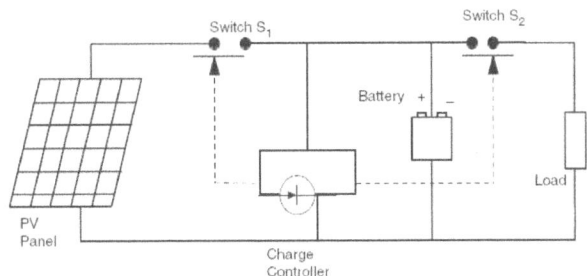

Source: Reference [11]

(Figure 2.7) Series charge regulator

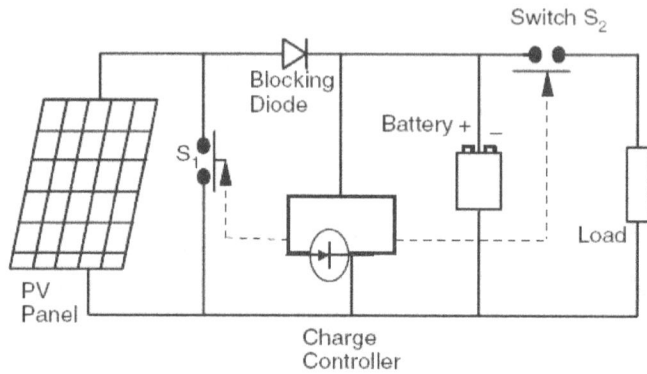

Source: Reference [11]
(Figure 2.8) Shunt charge regulator

2.1.4.2 DC–DC converter

Switch mode DC-to-DC converters are used to match the output of a PV generator to a variable load. There are various types of DC–DC converters such as:
- Buck (step-down) converter.
- Boost (step-up) converter.
- Buck–boost (step-down/up) converter.

(Figure 2.9) shows simplified diagrams of these three basic types converters. The basic concepts are an electronic switch, an inductor to store energy, and a "flywheel" diode, which carries the current during that part of switching cycle when the switch is off. The DC–DC converters allow the charge current to be reduced continuously in such a way that the resulting battery voltage is maintained at a specified value [11].

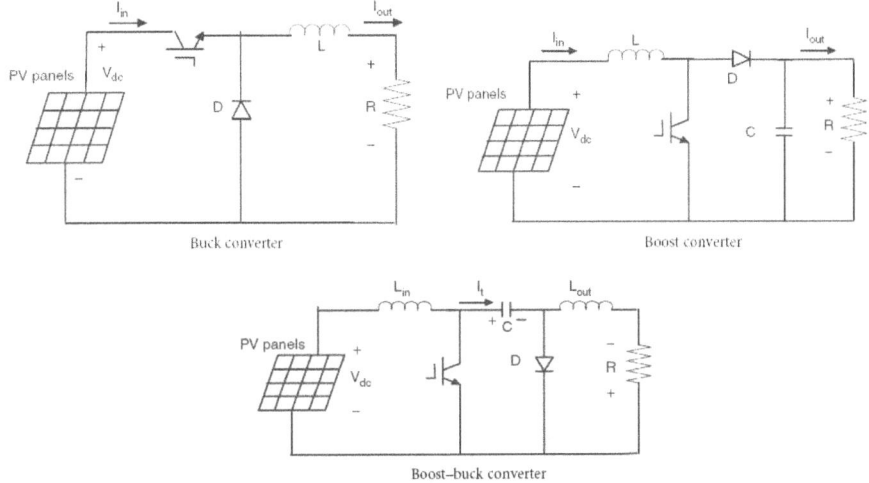

Source: Reference [11]
(Figure 2.9) DC-DC converters

2.1.4.3 Inverters

Inverters convert power from DC to AC while rectifiers convert it from AC to DC. Many inverters are bi-directional, i.e. they are able to operate in both inverting and rectifying modes.

In many stand-alone installations, alternating current is needed to operate 230V (or 110 V), 50 Hz (or 60 Hz) appliances. Generally stand-alone inverters operate at 12, 24, 48, 96, 120, or 240V DC depending upon the power level.

Several different semiconductor devices such as metal oxide semiconductor field effect transistor (MOSFETs) and insulated gate bipolar transistors (IGBTs) are used in the power stage of inverters. Typically, MOSFETs are used in units up to 5 kVA and 96V DC. They have the advantage of low switching losses at higher frequencies. Because the on-state voltage drop is 2V DC, IGBTs are generally used only above 96V DC systems.

Voltage source inverters are usually used in stand-alone applications. They can be single phase or three phase. There are three switching techniques commonly used: square wave, quasi-square wave, and pulse width modulation. Square-wave or modified square-wave inverters can supply power tools, resistive heaters, or incandescent lights, which do not require a high quality sine wave for reliable and efficient operation. However, many household appliances require low distortion sinusoidal waveforms. The use of true sine-wave inverters is recommended for remote area power systems. Pulse width modulated (PWM) switching is generally used for obtaining sinusoidal output from the inverters [11].

Design criteria and functions of PV inverters are as follows:
• Efficiency: well above 90% efficiency at 5% of nominal load
• Cost
• Weight and size: inverters with a 50-Hz transformer typically weigh more than 10 kg/kW, sometimes causing a problem with handling
• Voltage and current quality: harmonics and EMI

- Overload capability: approximately 20–30% for grid-connected inverters and up to 200% for short term overload for island inverters
- Precise and robust MPP tracking (reliably finding the overall MPP in partial shading situations)
- Supervision of the grid, safety/ENS (Einrichtung zur Netzüberwachung)
- Data acquisition and monitoring

To optimize inverters following these partially contradictory aims, the design of the power stages and control algorithms are increasingly more integrated. To implement suitable control schemes, digital signal processors are increasingly used. To account for the fact that PV electricity is yielded following a typical distribution, in Europe a weighted efficiency measure is used to aggregate the efficiency curve of inverters—the European efficiency. Good inverters in the kilowatt power range have a European efficiency of 92–96% and up to 97% for large central inverters. Inverters also perform MPP tracking in order to optimally operate the PV generator. Many different algorithms are used for MPP tracking for the following:

- Precision: high precision of MPP tracking requires high-precision measurement components
- Finding the global maximum power output in the case of partial shading, when a local maximum can occur
- Quickly adapting the MPP to changes in insolation (e.g., if clouds pass by) [9].

2.1.4.4 Wind turbine charge controllers

The basic block diagram of a stand-alone wind generator and battery charging system is shown in (Figure 2.10). The function of charge controller is to feed the power from the wind generator to the battery bank in a controlled manner. In the commonly used permanent magnet generators, this is usually done by using the controlled rectifiers. The controller should be designed to limit the maximum current into the battery, reduce charging current for high battery SOC, and maintain a trickle charge during full SOC periods [11].

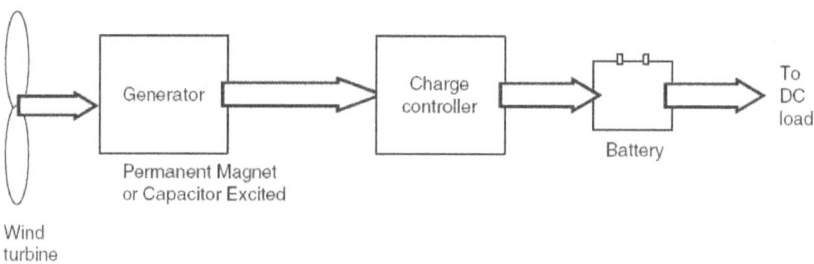

Source: Reference [11]
(Figure 2.10) Block diagram for a stand-alone wind generator and battery charging system

2.1.5 Backup generator

In the renewable energy systems, backup generators are used to provide additional electricity in times of low insolation or low wind speeds, to increase system availability, and to reduce overall electricity cost. In most medium or large power applications, diesel generators are the cheapest backup solution (backup capacity of 5 kW or more). In many countries, subsidized prices for diesel contribute significantly. However, the major disadvantages of piston-type backup generators are their need for maintenance every few hundred hours of operation and bad efficiency during operation at partial load, which is the common operation mode. In remote areas, access to the systems is often difficult or time-consuming, thus imposing extra cost for maintenance and fuel transport. The availability of spare parts and competent maintenance staff can be a major problem in developing countries in which diesel or PV/Wind–diesel hybrid systems are used for electrification of villages. A major problem regarding relatively small hybrid power systems is the lack of suitable backup generators in the power range of 1 kW or less. Thus, oversized piston-type generators are sometimes used for applications in which 1-kW backup capacity would be sufficient. Due to the poor efficiency of diesel engines under partial load conditions, the overall system efficiency of such systems is not satisfactory. In small hybrid systems, thermoelectric converters with a backup capacity of 10 to hundreds of watts are also used. In the medium term, fuel cells are an interesting alternative to small diesel or thermoelectric units because they promise high efficiency and low maintenance cost. However, the problem of hydrogen supply needs to be resolved [09].

2.2 Types of renewable energy systems
Renewable energy systems can be classified as:
- Stand-alone systems.
- Grid-connected systems.
- Hybrid systems.

2.2.1 Stand-alone systems
Stand-alone renewably (PV or Wind) powered systems (Figure 2.11) are used when there is no connection to an electricity grid. In order to ensure the supply of the stand-alone system with electric power also in the times without radiation (e.g., at night) or with very low radiation (e.g., at times with a strong cloud cover) or with low wind speeds, stand-alone systems mostly have an integrated storage system [4].

2.2.2 Grid-connected systems
The renewable source is connected here to a grid through inverters without battery storage (Figure 2.12). These systems can be classified as small systems like the residential rooftop systems or large grid-connected systems. The grid-interactive inverters must be synchronized with the grid in terms of voltage and frequency [11].

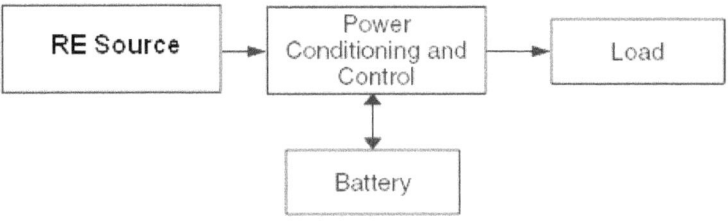

(Figure 2.11) Stand-alone system

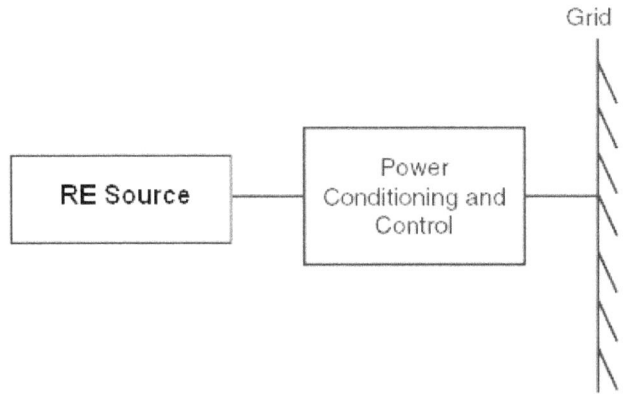

(Figure 2.12) Grid-connected system

2.2.3 Hybrid systems

The combination of RES, such as PV arrays or wind turbines, with engine-driven generators and battery storage, is widely recognized as a viable alternative to conventional remote area power supplies (RAPS). These systems are generally classified as hybrid energy systems (HES) (Figure 2.13).

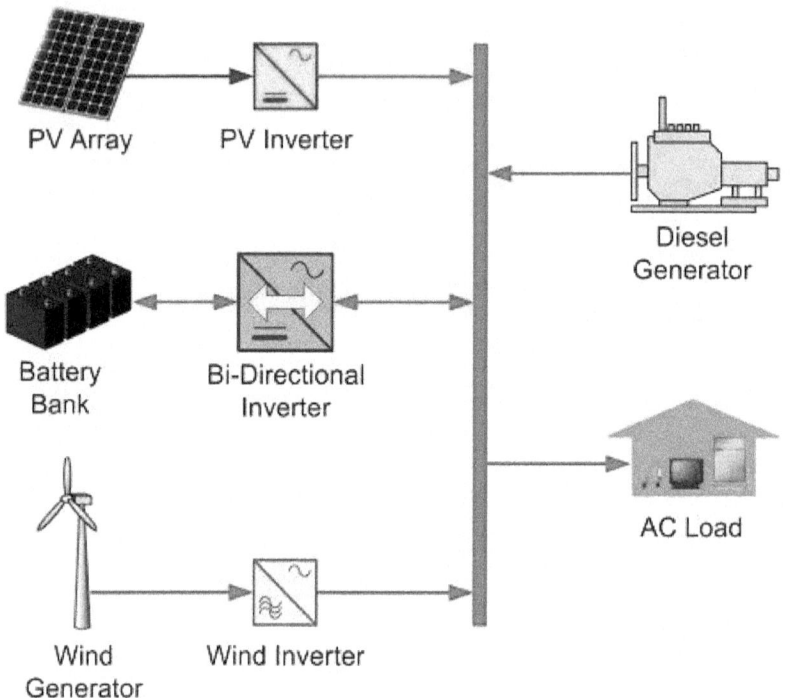

(Figure 2.13) A typical hybrid energy system

They are used increasingly for electrification in remote areas where the cost of grid extension is prohibitive and the price for fuel increases drastically with the remoteness of the location. For many applications, the combination of renewable and conventional energy sources compares favorably with fossil fuel-based RAPS systems, both in regard to their cost and technical performance. Because these systems employ two or

more different sources of energy, they enjoy a very high degree of reliability as compared to single-source systems such as a stand-alone diesel generator or a stand-alone PV or wind system. Applications of hybrid energy systems range from small power supplies for remote households, providing electricity for lighting and other essential electrical appliances, to village electrification for remote communities has been reported. Hybrid energy systems generate AC electricity by combining RES such as PV array with an inverter, which can operate alternately or in parallel with a conventional engine-driven generator. They can be classified according to their configuration as [11]:
- Series hybrid energy systems.
- Switched hybrid energy systems.
- Parallel hybrid energy systems.

The parallel hybrid systems can be further divided to DC or AC coupling.

2.2.3.1 Series configuration

In the conventional series hybrid systems shown in (Figure 2.14), all power generators feed DC power into a battery. Each component has therefore to be equipped with an individual charge controller and in the case of a diesel generator with a rectifier. To ensure reliable operation of series hybrid energy systems both the diesel generator and the inverter have to be sized to meet peak loads. This results in a typical system operation where a large fraction of the generated energy is passed through the battery bank, therefore resulting in increased cycling of the battery bank and reduced system efficiency. AC power delivered to the load is converted from DC to regulated AC by an inverter or a motor generator unit. The power generated by the diesel generator is first rectified and subsequently converted back to AC before being supplied to the load, which incurs significant conversion losses. The actual load demand determines the amount of electrical power delivered by the PV array, wind generator, the battery bank, or the diesel generator. The solar and wind charger prevents overcharging of the battery bank from the PV generator when the PV power exceeds the load demand and the batteries are fully charged. It may include MPPT to improve the utilization of the available PV energy, although the energy gain is marginal for a well-sized system. The system can be operated in manual or automatic mode, with the addition of appropriate battery voltage sensing and start/stop control of the engine-driven generator.

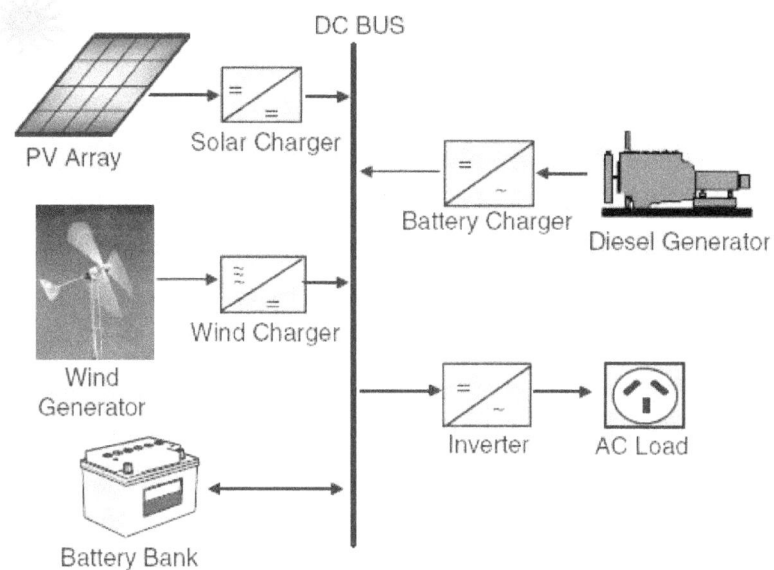

Source: Reference [11]

(Figure 2.14) Series Hybrid Energy System

Advantages:

• The engine-driven generator can be sized to be optimally loaded while supplying the load and charging the battery bank, until a battery SOC of 70–80% is reached.

• No switching of AC power between the different energy sources is required, which simplifies the electrical output interface.

• The power supplied to the load is not interrupted when the diesel generator is started.

- The inverter can generate a sine-wave, modified squarewave, or square-wave depending on the application.

Disadvantages:
- The inverter cannot operate in parallel with the engine-driven generator, therefore the inverter must be sized to supply the peak load of the system.
- The battery bank is cycled frequently, which shortens its lifetime.
- The cycling profile requires a large battery bank to limit the depth-of-discharge (DOD).
- The overall system efficiency is low, since the diesel cannot supply power directly to the load.
- Inverter failure results in complete loss of power to the load, unless the load can be supplied directly from the diesel generator for emergency purposes.

2.2.3.2 Switched configuration

Despite its operational limitations, the switched configuration remains one of the most common installations in some developing countries. It allows operation with either the engine-driven generator or the inverter as the AC source, yet no parallel operation of the main generation sources is possible. The diesel generator and the RES can charge the battery bank. The main advantage compared with the series system is that the load can be supplied directly by the engine-driven generator, which results in a higher overall conversion efficiency. Typically, the diesel generator power will exceed the load demand, with excess energy being used to recharge the battery bank. During periods of low electricity demand the diesel generator is switched off and the load is supplied from the PV array together with stored energy.

Switched hybrid energy systems can be operated in manual mode, although the increased complexity of the system makes it highly desirable to include an automatic controller, which can be implemented with the addition of appropriate battery voltage sensing and start/stop control of the engine-driven generator (Figure 2.15).

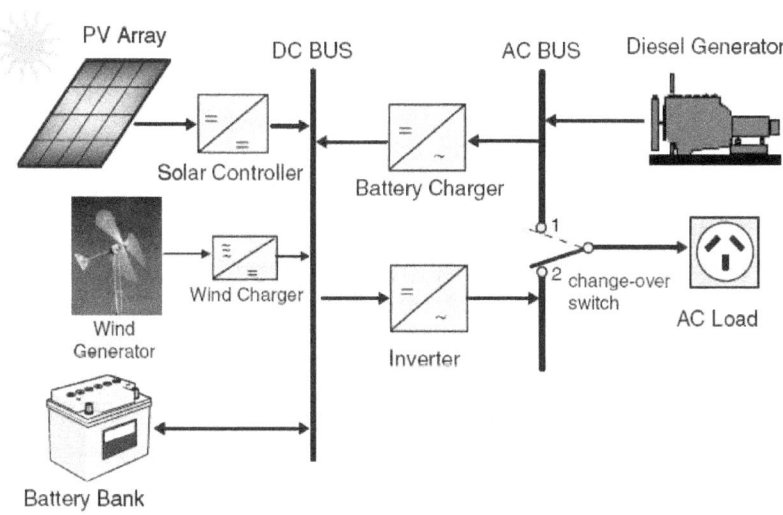

Source: Reference [11]

(Figure 2.15) Switched Hybrid Energy System

Advantages:
• The inverter can generate a sine-wave, modified squarewave, or square-wave, depending on the particular application.
• The diesel generator can supply the load directly, therefore improving the system efficiency and reducing the fuel consumption.

Disadvantages:
• Power to the load is interrupted momentarily when the AC power sources are transferred.
• The engine-driven alternator and inverter are typically designed to supply the peak load, which reduces their efficiency at part load operation.

2.2.3.3 Parallel configuration

The parallel hybrid system can be further classified as DC and AC couplings as shown in (Figure 2.16). In both schemes, a bi-directional inverter is used to link between the battery and an AC source (typically the output of a diesel generator). The bi-directional inverter can charge the battery bank (rectifier operation) when excess energy is available from the diesel generator or by the renewable sources, as well as act as a DC–AC converter (inverter operation). The bi-directional inverter may also provide "peak shaving" as part of a control strategy when the diesel engine is overloaded. In (Figure 2.16a), the renewable energy sources (RES) such as photovoltaic and wind are coupled on the DC side. DC integration of RES results in "custom" system solutions for individual supply cases requiring high costs for engineering, hardware, repair, and maintenance. Furthermore, power system expandability for covering needs of growing energy and power demand is also difficult. A better approach would be to integrate the RES on the AC side rather than on the DC side as shown in (Figure 2.16b). Parallel hybrid energy systems are characterized by two significant improvements over the series and switched system configuration. The inverter plus the diesel generator capacity rather than their individual component ratings limit the maximum load that can be supplied. Typically, this will lead to a doubling of the system capacity. The capability to synchronize the inverter with the diesel generator allows greater flexibility to optimize the operation of the system. Future systems should be sized with a reduced peak capacity of the diesel generator, which results in a higher fraction of directly used energy and hence higher system efficiencies. By using the same power electronic devices for both inverter and rectifier operation, the number of system components is minimized. Additionally, wiring and system installation costs are reduced through the integration of all power-conditioning devices in one central power unit. This highly integrated system concept has advantages over a more modular approach to system design, but it may prevent convenient system upgrades when the load demand increases. The parallel configuration offers a number of potential advantages over other system configurations. These objectives can only be met if the interactive operation of the individual components is controlled by an "intelligent" hybrid energy management system. Although today's generation of parallel systems include system controllers of varying complexity and sophistication, they do not optimize the performance of the complete system. Typically, both the diesel generator and the inverter are sized to supply anticipated peak loads. As a result, most parallel hybrid energy systems do not utilize their capability of parallel, synchronized operation of multiple power sources.

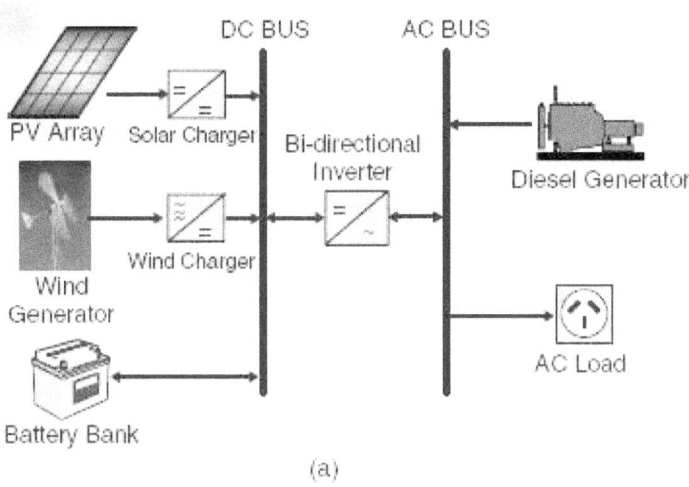

(a)

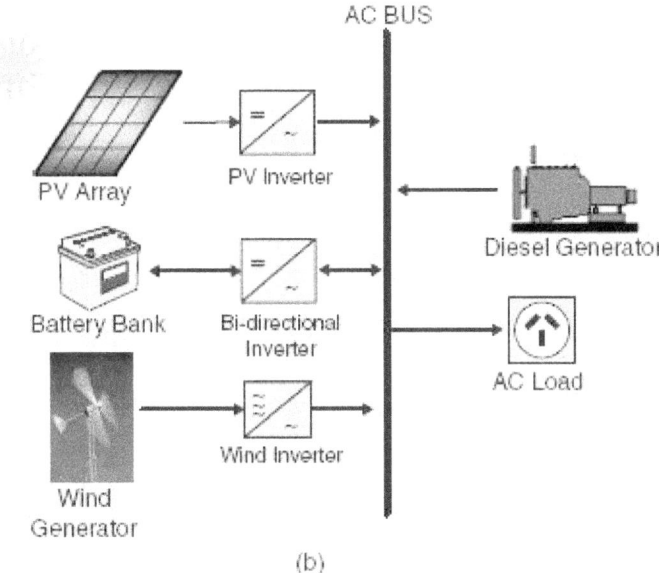

(b)

Source: Reference [11]

(Figure 2.16) Parallel PV-diesel hybrid energy system: (a) DC decoupling and (b) AC coupling

Advantages:
- The system load can be met in an optimal way.
- Diesel generator efficiency can be maximized.
- Diesel generator maintenance can be minimized.
- A reduction in the rated capacities of the diesel generator, battery bank, inverter, and renewable resources is feasible, while also meeting the peak loads.

Disadvantages:
- Automatic control is essential for the reliable operation of the system.
- The inverter has to be a true sine-wave inverter with the ability to synchronize with a secondary AC source.
- System operation is less transparent to the untrained user of the system [11].

Chapter 3: Sizing the hybrid system

3.1 Introduction

The Hybrid Energy System (HES) has received much attention over the past decade. It is a viable alternative solution as compared to systems, which rely entirely on hydrocarbon fuel. Apart from the mobility of the system, it also has longer life cycle. In particular, the integrated approach makes a hybrid system to be the most appropriate for isolated communities of a rural remote area [14].

For an ecologic and economic development, the coupling of a photovoltaic (PV)–wind system with battery storage may be very interesting when the local conditions are favorable to the level of wind and solar irradiation.
Because of the intermittent solar irradiation and wind speed characteristics, which highly influence the resulting energy production, the major aspects in the design of PV and wind generator (WG) power generation systems are the reliable power supply of the consumer under varying atmospheric conditions and the corresponding total system cost. Then it is essential to select the number of PV modules, WGs and batteries, and their installation details such that power is uninterruptedly supplied to the load and simultaneously the minimum system cost is achieved [15].

Designing of a hybrid energy system is site specific and it depends upon the resources available and the load demand.

The proposed system is designed to electrify a medium living compound in a remote area south of Khartoum the capital of Sudan, it will use the parallel configuration for the system resources as described in section (2.2.3.3) and it will be then simulated and its cost and reliability will be analyzed and optimized by the optimization and simulation tool HOMER® (See section 3.8).

3.2 Sun and Wind data and load demand profile
3.2.1 Sun and Wind data
(Table 3.1) and (Figure 3.1) and (Figure 3.2) showing the average global radiation [kWh/m²/d] and the average wind speed [m/s] for Khartoum city, Sudan (Latitude 15° 34' N, Longitude 32° 36' E)

Month	Global Radiation [kWh/m²/d]	Monthly average wind speed [m/s]
January	5.800	4.020
February	6.600	4.470
March	7.100	4.920
April	7.500	4.020
May	7.200	4.020
June	6.600	5.360
July	6.500	5.360
August	6.300	5.360
September	6.400	4.020
October	6.300	4.020
November	6.100	4.020
December	6.500	4.020
Average	**6.5**	**4.5**

Source: Reference [16]

(Table 3.1) Average global radiation and average wind speed for Khartoum

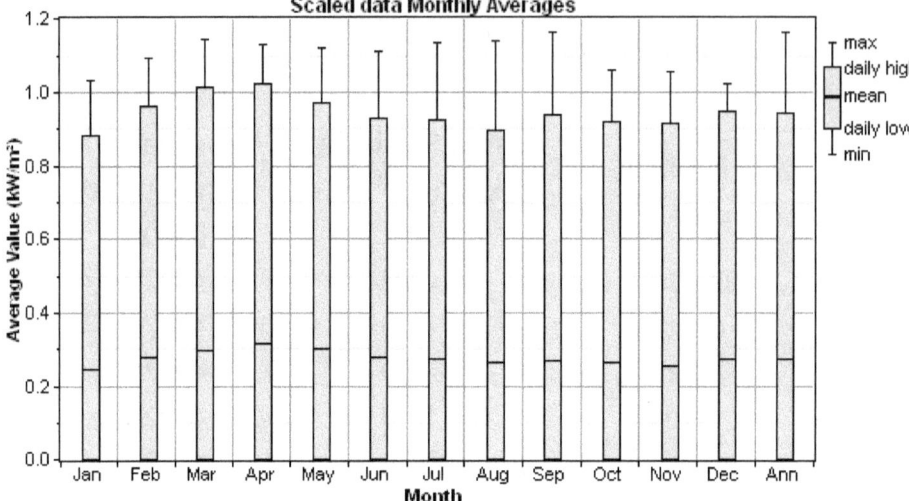

(Figure 3.1) Khartoum monthly global radiation average

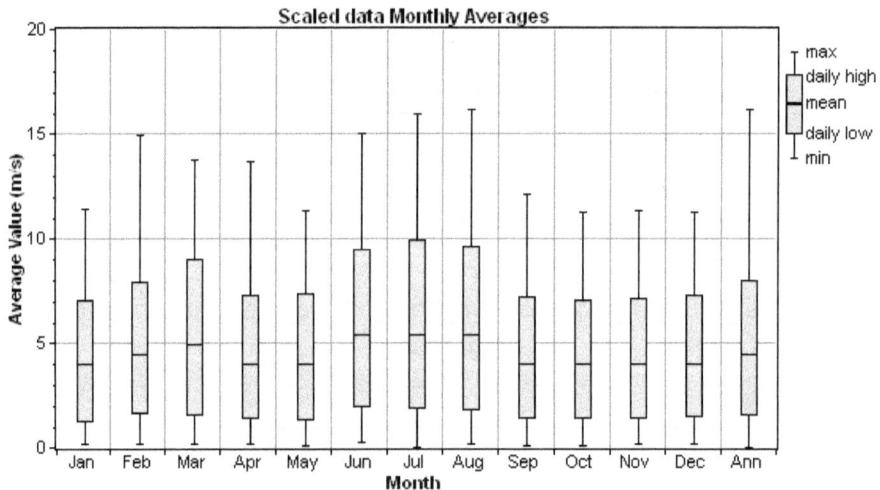

(Figure 3.2) Khartoum monthly wind speed average

3.2.2 Load demand profile

The load demand has been determined after excluding the unnecessary loads and replacing all the incandescent bulbs by fluorescent ones, because they're consuming less energy while giving the same required illumination. (Table 3.2) is showing the loads, their expected operating times and the total consumed power by the loads and (Figure 3.3) is for the daily load profile curve.

Load	Power [W]	Operating Time [h]	Consumed power per day [Wh/d]
Mosque loads			
Lighting	20x20	3	1200
Fans	90x60	5	2700
Sound system	90x1	2.25	202.5
Sleepers' loads			
Lighting	40x14	7	3920
Fans	90x14	13	16380
Bathrooms	20x10	4	800
Recreation room's loads			
Lighting	20x5	7	700
Fans	90x2	6	1080
TV	90x1	6	540
Water cooler	124.3x2	17	4226.2
Diner's lighting	40x1	4	1600
Clinic's loads			
Lighting	20x12	3	720
Fans	90x5	11	4950
Vaccine refrigerator	124.3x1	24	2983.2
Air conditioner	248.6x1	11	2734.6
Water cooler	124.3x1	17	2113.1
Lab appliances	50x2	11	1100
	5000		47967.6

(Table 3.2) Load profile table for the living compound

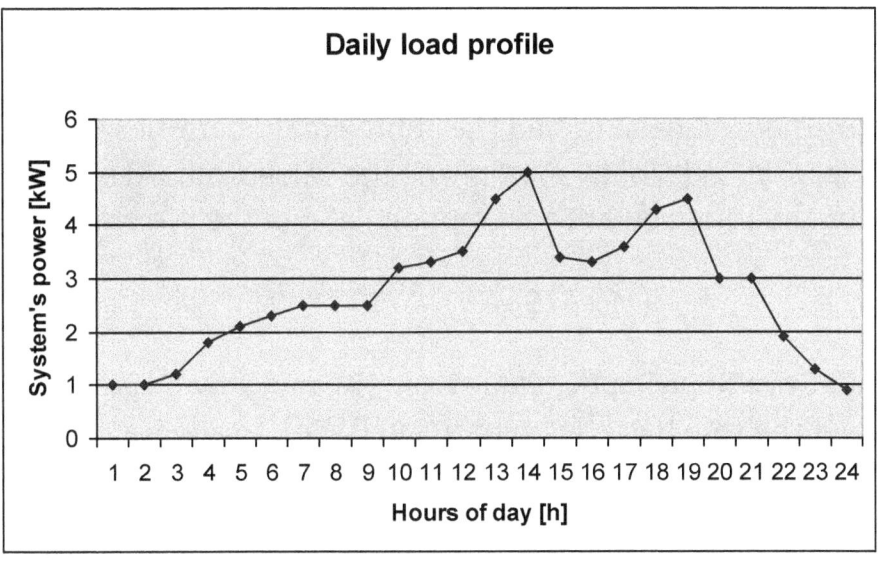

(Figure 3.3) Daily load profile

3.3 Inverter sizing

Sizing the inverter is easily achieved by finding the total power of the loads, which is in our case is 5 [kW], a good practice is to leave a room for future loads by adding 25% capacity to the inverter, hence:

Inverter capacity = 5 + (5 x 0.25) = 6.25 kW

3.4 Battery sizing

A) First, we have to find the number of "days of autonomy" (When we have to depend totally on the batteries; depends on the weather and other factors) and it is usually between 1-5 days, in this case I picked (1)

B) Then we have to find the approximate batteries capacity in [Wh] by multiplying the value from (A) above by the total power consumption [Wh] of the system:

$$(1) \times 47967.6 = 47967.6 \text{ [Wh]}$$

C) By multiplying the value from (B) above by (2) we will get the safe batteries capacity in [W/h], to allow a maximum discharge rate of 50% in normal conditions and another 50% in emergencies:

$$47967.6 \times 2 = 95935.2 \text{ [Wh]}$$

D) Then we have convert the value from (C) above from [Wh] to [Ah] by dividing it by the expected system voltage (12 or 24 or 48 V_{DC}), I picked 24 V_{DC}:

$$95935.2/24 = 3997.3 \text{ Ah}$$

E) Next we have to choose the type of battery we will use for our system, the selected battery was BP 6P363, Voltage = 6 V, Capacity = 363 Ah

F) The number of rows of parallel-connected batteries are determined by dividing the result from step (D) above by the capacity of the single battery [Ah]:

$$3997.3/363 = 12 \text{ rows}$$

G) While the number of the serial-connected batteries is determined by dividing the system voltage by the battery's voltage:

$$24/6 = 4 \text{ batteries}$$

H) The total number of batteries for the system is the product of (F) and (H):

$$12 \times 4 = 48 \text{ batteries}$$

3.5 Determining the number of parallel and serial PV modules

To build a required 3 [kW] PV array for this system, I picked the PV module KC80 from Kyocera Solar Company as a core for my PV generator, the module has the specifications in (Table 3.3) under the standard testing conditions (STC):

■ Electrical Specifications

MODEL	KC80
Maximum Power	80 Watts
Maximum Power Voltage	16.9 Volts
Maximum Power Current	4.73 Amps
Open Circuit Voltage	21.5 Volts
Short-Circuit Current	4.97 Amps
Length	976mm (38.4in.)
Width	652mm (25.7in.)
Depth	52mm (2.0in.)
Weight	9.6kg (21.2lbs.)

Note: The electrical specifications are under test conditions of Irradiance of 1kW/m², Spectrum of 1.5 air mass and cell temperature of 25°C

(Table 3.3) Electrical specifications of KC80 PV module

Then I followed the following steps to determine the number of PV modules connected in serial and the number of rows for the ones connected in parallel:

A) The number of PV modules connected in serial (N_S) is obtained by dividing the system DC voltage by the operating voltage (V_{PMAX}) for one module (From Table 3.3):

$$N_S = V_{DC}/V_{PMAX}$$

$$N_S = 24/16.9 = 2$$

B) The number of rows of PV modules connected in parallel can be obtained after finding the load current I_L and the PV current I_{PV}:

$$I_L = E_L/24\ V_{DC}\ [A]$$

Where $E_L \equiv$ The consumed power by loads [Wh/d]

$24 \equiv$ hours of a day

This value is then multiplied by 0.29, which is the ratio of PV generation contribution to the system (Obtained from HOMER simulation results):

$$I_L = (0.29 \times 47967.6) / (24 \times 24) = 24.150\ [A]$$

$$I_{PV} = 24 \times I_L/PSH\ [A]$$

Where PSH $\equiv$ Peak Solar Hours, equals to the average global radiation [kWh/m²/d] = 6.5 (From Table 3.1)

$$I_{PV} = (24 \times 24.150) / 6.5 = 89.171\ [A]$$

$$N_P = I_{PV}/I_{SC} = 89.171/4.97 = 18$$

Where $N_P \equiv$ Number of in-parallel PV rows

$I_{SC} \equiv$ The PV module Short-Circuit current (From Table 3.3)

Thus, the total number of PV modules = $N_S \times N_P$

$$2 \times 18 = 36\ \text{module}$$

3.6 Sizing the charge regulator

To size the charge regulator we have to take into consideration that it should handle the maximum current produced by the PV array and the maximum load current if it has the load control feature, also some other factors (like the reflected rays from the buildings' roofs for example) might increase the PV array current by 25% above the rated Short-Circuit current of the modules, thus the size of the regulator is determined by the sum of the Short-Circuit currents for the in-parallel modules' rows:

$$4.97 \times 23 = 114.31$$

Some standards like UL and NEC recommend increasing the regulator capacity by 56%, thus the final regulator size will be:

$$114.31 + (114.31 \times 0.56) = 178.3 \, [A]$$

As a result, a regulator with 179 [A] capacity will do the job.

3.7 Choosing the wind turbine and the generator

I picked the Whisper 200 Wind Turbine (formerly H80) by Southwest Windpower as a WG for the system, (Table 3.4) is showing the turbine specifications:

Whisper 200 Technical Specifications

WHISPER 200

Rated Power	1000 watts at 11.6 m/s (26 mph)
Monthly Energy	200 kWh/mo at 5.4 m/s (12 mph)
Start-Up Wind Speed	3.1 m/s (7 mph)
Rotor Diameter	2.7 m (9 ft)
Voltage	12, 24, 48 VDC*; HV Available at 120v, 230v
Overspeed Protection	Patented side-furling
Turbine Controller	Whisper controller (Optional with all Units)
Mount	6.35 cm pipe (2.5 in schedule 40)
Body	Cast aluminum with corrosion resistant finish
Blades	(3) Carbon reinforced fiberglass
Survival Wind Speed	55 m/s (120 mph)
Weight	30 kg (65 lb) box: 39.46 kg (87 lb)
Shipping Dimensions	1295 x 508 x 330 mm (51 x 20 x 13 in)
Warranty	5 year limited warranty

*Power ratings are normalized for sea level.

(Table 3.3) Whisper 200 Wind Turbine specifications

A standard synchronous diesel engine with capacity of 3 kW was picked for the system.

3.8 HOMER simulation and optimization tool
3.8.1 What's HOMER?

HOMER, the micropower optimization model, simplifies the task of evaluating designs of both off-grid and grid-connected power systems for a variety of applications. When you design a power system, you must make many decisions about the configuration of the system: What components does it make sense to include in the system design? How many and what size of each component should you use? The large number of technology options and the variation in technology costs and availability of energy resources make these decisions difficult. HOMER's optimization and sensitivity analysis algorithms make it easier to evaluate the many possible system configurations.

To use HOMER, you provide the model with inputs, which describe technology options, component costs, and resource availability. HOMER uses these inputs to simulate different system configurations, or combinations of components, and generates results that you can view as a list of feasible configurations sorted by net present cost. HOMER also displays simulation results in a wide variety of tables and graphs that help you compare configurations and evaluate them on their economic and technical merits. You can export the tables and graphs for use in reports and presentations. (Figure 3.4) for HOMER main screen.

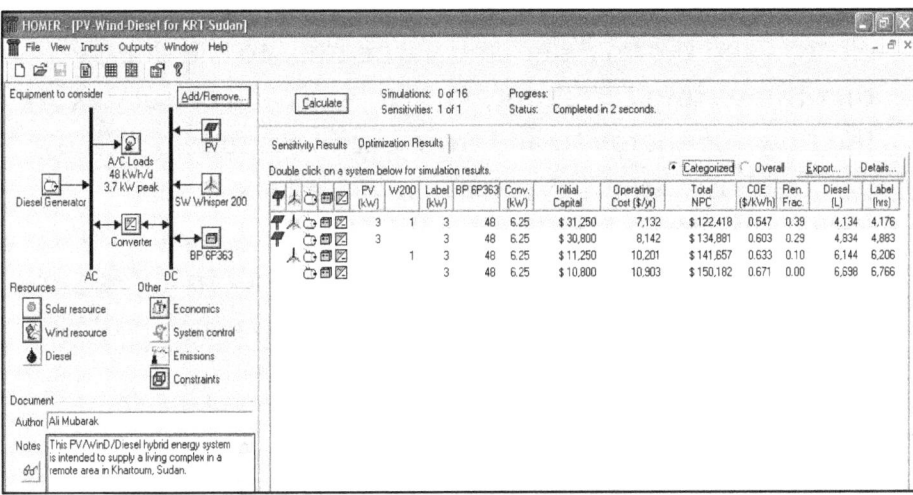

(Figure 3.4) HOMER main screen

When you want to explore the effect that changes in factors such as resource availability and economic conditions might have on the cost-effectiveness of different system configurations, you can use the model to perform sensitivity analyses. To perform a sensitivity analysis, you provide HOMER with sensitivity values that describe a range of resource availability and component costs. HOMER simulates each system configuration over the range of values. You can use the results of a sensitivity analysis to identify the factors that have the greatest impact on the design and operation of a power system. You can also use HOMER sensitivity analysis results to answer general questions about technology options to inform planning and policy decisions.

HOMER simulates the operation of a system by making energy balance calculations for each of the 8,760 hours in a year. For each hour, HOMER compares the electric and thermal demand in the hour to the energy that the system can supply in that hour, and calculates the flows of energy to and from each component of the system. For systems that include batteries or fuel-powered generators, HOMER also decides for each hour how to operate the generators and whether to charge or discharge the batteries.

HOMER performs these energy balance calculations for each system configuration that you want to consider. It then determines whether a configuration is feasible, i.e., whether it can meet the electric demand under the conditions that you specify, and estimates the cost of installing and operating the system over the lifetime of the project. The system cost calculations account for costs such as capital, replacement, operation and maintenance, fuel, and interest.

After simulating all of the possible system configurations, HOMER displays a list of configurations, sorted by net present cost (sometimes called lifecycle cost), that you can use to compare system design options.

When you define sensitivity variables as inputs, HOMER repeats the optimization process for each sensitivity variable that you specify. For example, if you define wind speed as a sensitivity variable, HOMER will simulate system configurations for the range of wind speeds that you specify.

The HOMER website, www.nrel.gov/homer, contains the latest information on the model, as well as sample files, resource data, and contact information [17].

3.8.2 Program Inputs
To start simulation, we need to provide HOMER with the data of solar radiation, wind speeds and the load demand profile in addition to system components and their prices. Data of PV array and wind turbine are entered under "Resources" and "Components" fields while the other system components are entered under "Components" field.

3.8.2.1 Solar and PV inputs
(Table 3.1) was used to generate the input data for the solar resource and the PV inputs entered base on the selected Kyocera KC80 PV module.

Latitude
We have to specify the latitude and the amount of solar radiation available to the photovoltaic (PV) array throughout the year. The latitude specifies your location on the Earth's surface. It is an important variable in solar calculations. It is used when calculating radiation values

from clearness indices, and vice versa. It is also used to calculate the radiation incident on a tilted surface. HOMER uses this data to calculate the output of the PV array each hour of the year.

Baseline data

The baseline data is the set of 8,760 values representing the average global solar radiation on the horizontal surface, expressed in kWh/m2, for each hour of the year. HOMER displays the monthly average radiation and clearness index of the baseline data in the solar resource table and graph.

There are two ways to create baseline data: you can use HOMER to synthesize data, or you can import hourly radiation data from a file.

To synthesize data, you must enter twelve average monthly values of either solar radiation or clearness index. You do not have to enter both; HOMER calculates one from the other using the latitude. Enter each monthly value in the appropriate row and column of the solar resource table. As you enter values in the table, HOMER builds a set of 8,760 solar radiation values, or one for each hour of the year. HOMER creates the synthesized values using the Graham algorithm, which results in a data sequence that has realistic day-to-day and hour-to-hour variability and autocorrelation.

PV Properties

Variable	Description
Output current	Whether the PV array produces AC or DC power. All PV cells produce DC electricity, but some PV arrays have built-in inverters to convert to AC.
Lifetime	The number of years the PV panels will last.
Derating factor	A scaling factor applied to the PV array power output to account for reduced output in real-world operating conditions compared to operating conditions at which the array was rated.
Slope	The angle at which the panels are mounted relative to the horizontal.
Azimuth	The direction towards which the panels face.
Ground reflectance	The fraction of solar radiation incident on the ground that is reflected.
Tracking system	The type of tracking system used to direct the PV panels towards the sun.
Temperature coefficient of power	A number indicating how strongly the power output of the PV array depends on cell temperature.
Nominal operating cell temperature	The cell temperature at 0.8 kW/m2, 20°C ambient temperature, and 1 m/s wind speed.
Efficiency at standard test conditions	The maximum power point efficiency under standard test conditions.

(Table 3.4) HOMER PV properties

3.8.2.2 Wind and turbine inputs

(Table 3.1) was used to generate the input data for the wind resource and the Wind turbine inputs entered base on the selected Whisper 200 Wind Turbine.

Wind parameters

Variable	Description
Altitude	The altitude in meters above sea level.
Anemometer height	The height above ground at which the wind speed data were measured.
Weibull k	A measure of the long-term distribution of wind speeds.
Autocorrelation factor	A measure of the hour-to-hour randomness of the wind speed.
Diurnal pattern strength	A measure of how strongly the wind speed depends on the time of day.
Hour of peak wind speed	The time of day that tends to be windiest on average.

(Table 3.5) HOMER Wind parameters

3.8.2.3 Diesel generator inputs

System designers commonly specify just a single nonzero generator size, one large enough to comfortably serve the peak load. When given a choice of generator sizes, HOMER will invariably choose the smallest one that meets the maximum annual capacity shortage constraint, since smaller generators typically cost less to operate than larger generators.

Generator Properties

Variable	Description
Lifetime	The number of hours the generator will run before needing replacement.
Minimum load ratio	The minimum allowable load on the generator, as a percentage of its rated capacity.
Intercept coefficient	the no-load fuel consumption of the generator divided by its rated capacity
Slope	marginal fuel consumption of the generator
Substitution ratio	The ratio with which the biogas replaces fossil fuel in the cofired generator
Minimum fossil fraction	the minimum allowable fossil fraction
Derating factor	The relative capacity of the generator at the minimum fossil fraction.

(Table 3.6) HOMER Generator properties

3.8.2.4 Load profile

(Table 3.2) was used to generate the load demand profile in HOMER. This load profile was then entered under the "Primary load", which is the electrical load that must be met immediately in order to avoid unmet load. Each hour of the year, HOMER dispatches the power-producing components of the system to serve the total primary load.

3.8.3 Program outputs and results

After running the simulation, we were able to find the optimum design in our specific case and HOMER generated a detailed report for each operating hour in the year. (Table 3.7) showing the optimized day 1 simulation results.

Date	End Time	Solar Radiation (kW/m2)	Incident Solar (kW/m2)	Wind Speed (m/s)	AC Prim. Load (kW)	PV Power (kW)	W200 Power (kW)	Label Power (kW)	AC Prim. Served (kW)	Inverter Input (kW)	Inverter Output (kW)	Rectifier Input (kW)	Rectifier Output (kW)	Battery Input (kW)	Battery SOC (%)	Battery Energy ($/kWh)
Jan 1	1:00	0.00	0.00	3.1	0.73	0.00	0.03	0.00	0.73	0.81	0.73	0.00	0.00	-0.79	99.1	0.000
Jan 1	2:00	0.00	0.00	2.6	0.73	0.00	0.00	0.00	0.73	0.81	0.73	0.00	0.00	-0.81	98.1	0.000
Jan 1	3:00	0.00	0.00	2.4	0.88	0.00	0.00	0.00	0.88	0.97	0.88	0.00	0.00	-0.97	97.0	0.000
Jan 1	4:00	0.00	0.00	0.8	1.32	0.00	0.00	0.00	1.32	1.46	1.32	0.00	0.00	-1.46	95.3	0.000
Jan 1	5:00	0.00	0.00	1.1	1.54	0.00	0.00	0.00	1.54	1.71	1.54	0.00	0.00	-1.71	93.3	0.000
Jan 1	6:00	0.00	0.00	1.2	1.68	0.00	0.00	0.00	1.68	1.87	1.68	0.00	0.00	-1.87	91.1	0.000
Jan 1	7:00	0.04	0.07	1.1	1.83	0.18	0.00	0.00	1.83	2.03	1.83	0.00	0.00	-1.86	88.9	0.000
Jan 1	8:00	0.17	0.20	1.8	1.83	0.49	0.00	0.00	1.83	2.03	1.83	0.00	0.00	-1.54	87.1	0.000
Jan 1	9:00	0.30	0.33	1.8	1.83	0.80	0.00	0.00	1.83	2.03	1.83	0.00	0.00	-1.23	85.7	0.000
Jan 1	10:00	0.51	0.58	1.6	2.34	1.39	0.00	0.00	2.34	2.60	2.34	0.00	0.00	-1.21	84.3	0.000
Jan 1	11:00	0.22	0.22	2.5	2.41	0.52	0.00	0.00	2.41	2.68	2.41	0.00	0.00	-2.16	81.7	0.000
Jan 1	12:00	0.19	0.19	1.3	2.56	0.46	0.00	0.00	2.56	2.84	2.56	0.00	0.00	-2.39	78.9	0.000
Jan 1	13:00	0.54	0.58	1.1	3.29	1.39	0.00	0.00	3.29	3.66	3.29	0.00	0.00	-2.26	76.3	0.000
Jan 1	14:00	0.46	0.49	1.7	3.66	1.17	0.00	0.00	3.66	4.06	3.66	0.00	0.00	-2.90	72.9	0.000
Jan 1	15:00	0.22	0.22	1.3	2.49	0.53	0.00	0.00	2.49	2.76	2.49	0.00	0.00	-2.23	70.3	0.000
Jan 1	16:00	0.11	0.11	2.9	2.41	0.27	0.01	0.00	2.41	2.68	2.41	0.00	0.00	-2.40	67.5	0.000
Jan 1	17:00	0.11	0.12	3.4	2.63	0.28	0.05	0.00	2.63	2.92	2.63	0.00	0.00	-2.60	64.4	0.000
Jan 1	18:00	0.01	0.00	3.3	3.14	0.00	0.04	0.00	3.14	3.49	3.14	0.00	0.00	-3.45	60.4	0.000
Jan 1	19:00	0.00	0.00	2.1	3.29	0.00	0.00	0.00	3.29	3.66	3.29	0.00	0.00	-3.66	56.1	0.000
Jan 1	20:00	0.00	0.00	3.3	2.19	0.00	0.04	0.00	2.19	2.44	2.19	0.00	0.00	-2.40	53.3	0.000
Jan 1	21:00	0.00	0.00	4.1	2.19	0.00	0.11	0.00	2.19	2.44	2.19	0.00	0.00	-2.33	50.6	0.000
Jan 1	22:00	0.00	0.00	3.4	1.39	0.00	0.05	0.00	1.39	1.54	1.39	0.00	0.00	-1.50	48.8	0.000
Jan 1	23:00	0.00	0.00	2.1	0.95	0.00	0.00	0.00	0.95	1.06	0.95	0.00	0.00	-1.06	47.6	0.000

(Table 3.7) HOMER Day 1 simulation results

The following detailed report has been generated for the optimum design:

System architecture

PV Array	3 kW
Wind turbine	1 SW Whisper 200
Diesel Generator	3 kW
Battery	48 BP 6P363
Inverter	6.25 kW
Rectifier	6.25 kW
Dispatch strategy	Cycle Charging

Cost summary

Total net present cost	$ 122,418
Levelized cost of energy	$ 0.547/kWh
Operating cost	$ 7,132/yr

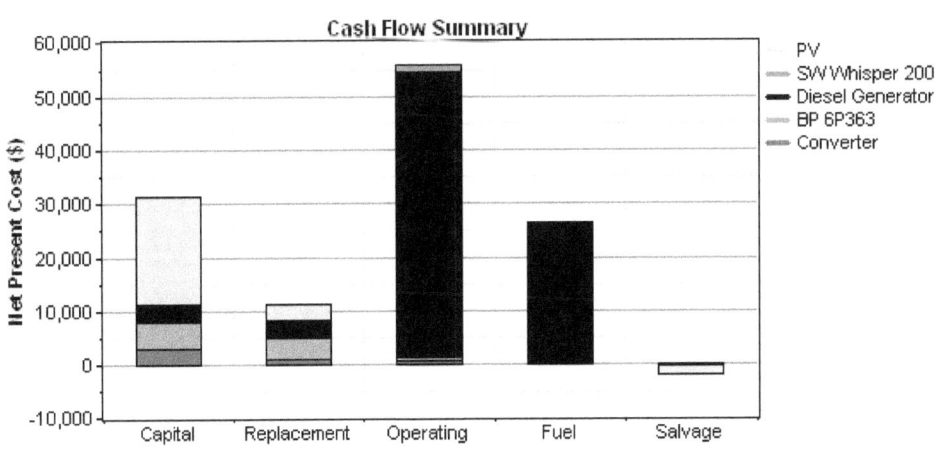

Net Present Costs

Component	Capital ($)	Replace-ment ($)	O&M ($)	Fuel ($)	Salvage ($)	Total ($)
PV	20,000	3,118	0	0	-1,747	21,371
SW Whisper 200	450	83	1,278	0	-16	1,796
Diesel Generator	3,000	3,072	53,383	26,422	-9	85,867
BP 6P363	4,800	4,053	511	0	-299	9,065
Converter	3,000	835	639	0	-155	4,318
System	31,250	11,161	55,812	26,422	-2,227	122,418

Annualized Costs

Component	Capital ($/yr)	Replace-ment ($/yr)	O&M ($/yr)	Fuel ($/yr)	Salvage ($/yr)	Total ($/yr)
PV	1,565	244	0	0	-137	1,672
SW Whisper 200	35	7	100	0	-1	141
Diesel Generator	235	240	4,176	2,067	-1	6,717
BP 6P363	375	317	40	0	-23	709
Converter	235	65	50	0	-12	338
System	2,445	873	4,366	2,067	-174	9,576

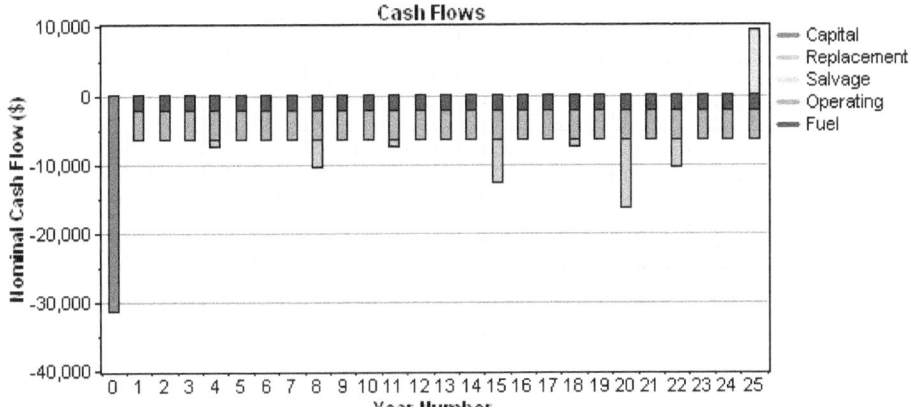

Electrical

Component	Production (kWh/yr)	Fraction
PV array	6,019	29%
Wind turbine	2,110	10%
Diesel Generator	12,526	61%
Total	20,655	100%

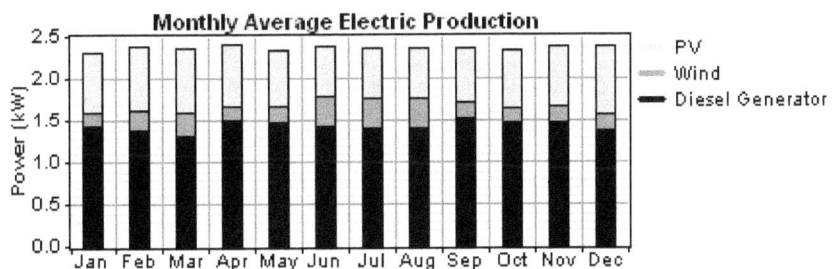

Load	Consumption (kWh/yr)	Fraction
AC primary load	17,508	100%
Total	17,508	100%

Quantity	Value	Units
Excess electricity	0.00541	kWh/yr
Unmet load	0.0000270	kWh/yr
Capacity shortage	0.363	kWh/yr
Renewable fraction	0.394	kWh/yr

PV

Quantity	Value	Units
Rated capacity	3.00	kW
Mean output	0.687	kW
Mean output	16.5	kWh/d
Capacity factor	22.9	%
Total production	6,019	kWh/yr
Minimum output	0.00	kW
Maximum output	2.89	kW
PV penetration	34.4	%
Hours of operation	4,360	hr/yr
Levelized cost	0.278	$/kWh

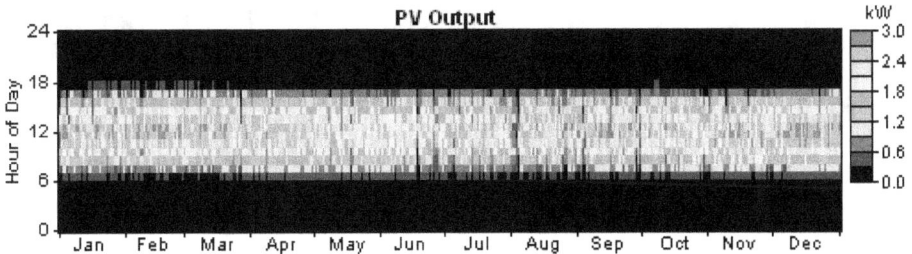

Wind Turbine: SW Whisper 200

Variable	Value	Units
Total rated capacity	1.00	kW
Mean output	0.241	kW
Capacity factor	24.1	%
Total production	2,110	kWh/yr
Minimum output	0.00	kW
Maximum output	1.00	kW
Wind penetration	12.1	%
Hours of operation	6,700	hr/yr
Levelized cost	0.0666	$/kWh

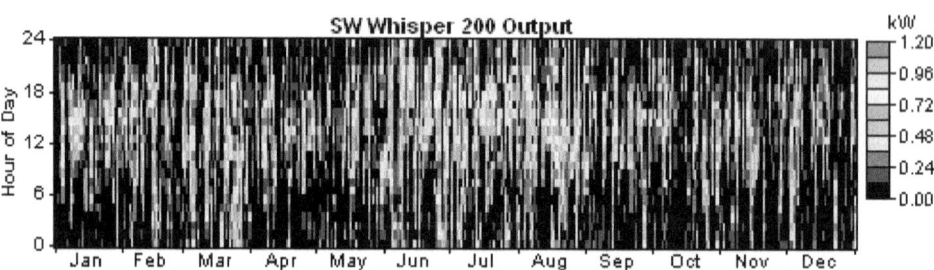

Diesel Generator

Quantity	Value	Units
Hours of operation	4,176	hr/yr
Number of starts	173	starts/yr
Operational life	3.59	yr
Capacity factor	47.7	%
Fixed generation cost	1.19	$/hr
Marginal generation cost	0.125	$/kWhyr
Electrical production	12,526	kWh/yr
Mean electrical output	3.00	kW
Min. electrical output	1.20	kW
Max. electrical output	3.00	kW
Fuel consumption	4,134	L/yr
Specific fuel consumption	0.330	L/kWh
Fuel energy input	40,676	kWh/yr
Mean electrical efficiency	30.8	%

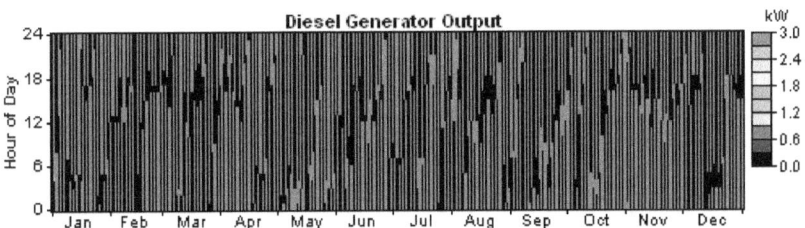

Quantity	Value
String size	1
Strings in parallel	48
Batteries	48
Bus voltage (V)	6

Quantity	Value	Units
Nominal capacity	105	kWh
Usable nominal capacity	62.7	kWh
Autonomy	31.4	hr
Lifetime throughput	46,272	kWh
Battery wear cost	0.070	$/kWh
Average energy cost	0.134	$/kWh

Quantity	Value	Units
Energy in	7,319	kWh/yr
Energy out	5,889	kWh/yr
Storage depletion	41.3	kWh/yr
Losses	1,389	kWh/yr
Annual throughput	6,584	kWh/yr
Expected life	7.03	yr

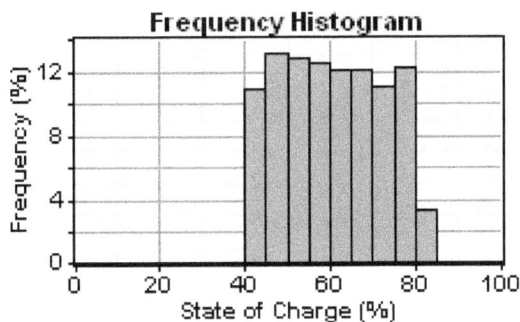

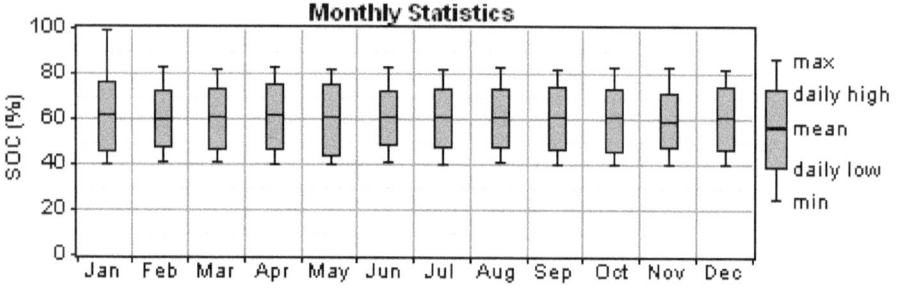

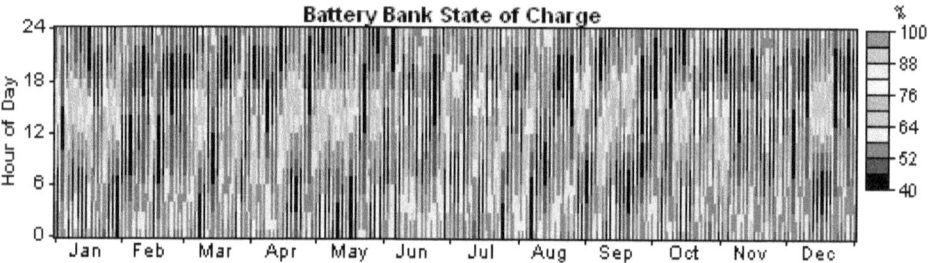

Converter

Quantity	Inverter	Rectifier	Units
Capacity	6.25	6.25	kW
Mean output	1.08	0.43	kW
Minimum output	0.00	0.00	kW
Maximum output	3.66	1.99	kW
Capacity factor	17.2	6.9	%
Hours of operation	5,270	3,490	hrs/yr
Energy in	10,487	4,456	kWh/yr
Energy out	9,438	3,788	kWh/yr
Losses	1,049	669	kWh/yr

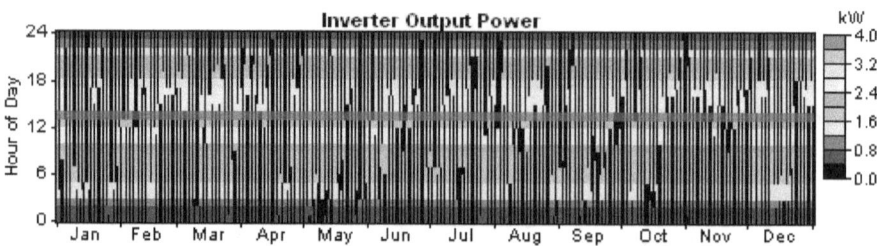

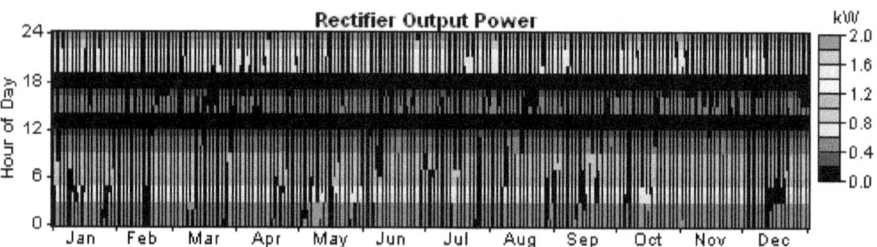

Emissions

Pollutant	Emissions (kg/yr)
Carbon dioxide	10,886
Carbon monoxide	26.9
Unburned hydrocarbons	2.98
Particulate matter	2.03
Sulfur dioxide	21.9
Nitrogen oxides	240

Chapter 4: Conclusion

The utilization of renewable sources of energy such as wind and solar power has experienced rapid growth in the past decade, and most of them are pollution-free sources of abundant power. They may also eliminate the need for running new high-voltage transmission lines, which may involve a significant investment.

In this design the initial system cost was 31,250 $ which considered a small fraction from the cost of erecting the HV transmission lines which according to Sudanese Ministry of Energy and Mining might cost 30,000 $ for each 1 Km, while some rural communities in remote areas can be a hundreds kilometers away from the national grid. The huge cost of building a distribution substation should be also considered.

The information about local wind and solar indicates that a feasible hybrid energy system can be planned, modeled and designed for the above purpose.

In such designs, adequacy assessment in power system design with renewable distributed generations should be systematically addressed. This is especially necessary in order to adapt to the new design requirements with the increasing penetration of fluctuating renewable sources of energy. Uncertainty factors, such as generator failures and renewable power availability, need also to be taken into account in calculating system reliability indices with penetration of time-dependent sources.

Although, generally speaking, distributed generation using renewable resources accounts for a small proportion of the world's existing power supply thus far, it is anticipated that renewables will contribute more significantly in the near future.

References

[01] Wang, Lingfeng; Singh, Chanan, Hybrid Design of Electric Power Generation Systems Including Renewable Sources of Energy, SAGE Publications, Bulletin of Science, Technology & Society, Vol. 28, No. 3, 2008

[02] Sen, Zekai, Solar energy fundamentals and modeling techniques: atmosphere, environment, climate change and renewable energy, Springer-Verlag London Limited, 2008

[03] Kreith, Frank; Goswami, D. Yogi, Handbook of Energy Efficiency and Renewable Energy, CRC Press, 2007

[04] Goetzberger, Adolf; Hoffmann, Volker, Photovoltaic Solar Energy Generation, Springer-Verlag Berlin Heidelberg, 2005

[05] Mustafa Omer, Abdeen, Sudan's Renewable Energy Options: Power for Water and Living, Stockholm Environment Institute, Renewable Energy for Development, Vol. 13, No. 2, June 2000, pp. 4-5

[06] Abd El Gadir El Sheikh, Kawthar, Current Energy and Environment Issues in Sudan, The National Conference on The Development & Environment, Sudan, Sep-2006

[07] Global Environment Facility (GEF) website www.gefweb.org

[08] Introduction to Photovoltaic Systems, SECO Fact Sheet No. 11, Texas State Energy Conservation Office, www.InfinitePower.org

[09] Cleveland, Cutler J. et al., Encyclopedia of Energy, Volume V, Elsevier Science & Technology Books, 2004

[10] Vanek, Francis M.; Albright, Louis D., Energy Systems Engineering: Evaluation and Implementation, McGraw-Hill, 2008

[11] Rashid, Muhammad H., Power Electronics Handbook, Second Edition, Academic Press, 2006

[12] Markvart, Tomas; Castaner, Luis, Practical handbook of photovoltaics: fundamentals and applications, Elsevier Advanced Technology, 2003

[13] Nayar, C. et. al, A case study of a PV/wind/diesel hybrid energy system for remote islands in the republic of Maldives, Universities Power Engineering Conference, 2007. AUPEC 2007. Australasian, 12-9 Dec-2007, pp. 1-7

[14] Gupta, A.; Saini, R.P.; Sharma, M.P., Design of an Optimal Hybrid Energy System Model for Remote Rural Area Power Generation, International Conference on

Electrical Engineering, 2007 (ICEE '07), 11-12 Apr-2007, pp. 1-6

[15] Belfkira, R.; Hajji, O.; Nichita, C.; Barakat, G., Optimal sizing of stand-alone hybrid wind/PV system with battery storage, European Conference on Power Electronics and Applications, 2-5 Sep-2007, pp. 1-10

[16] Mubarak, A. et. al, Optimal Design of a PV/Wind/Diesel hybrid energy system, Diploma thesis, Sudan University of Science and Technology, Nov-2002

[17] Getting Started Guide for HOMER Version 2.1, National Renewable Energy Laboratory, Apr-2005

www.ingramcontent.com/pod-product-compliance
Lightning Source LLC
Chambersburg PA
CBHW072216170526
45158CB00002BA/623